Radiology

Radiology

Bradford J. Wood, MD

Clinical Assistant
Division of Abdominal and Interventional Radiology
Massachusetts General Hospital
Harvard Medical School
Boston, Massachusetts

Sangeeta Desai Wood, MD

Assistant in Emergency Medicine
Massachusetts General Hospital
Harvard Medical School
Boston, Massachusetts

Williams & Wilkins
A WAVERLY COMPANY

BALTIMORE • PHILADELPHIA • LONDON • PARIS • BANGKOK
BUENOS AIRES • HONG KONG • MUNICH • SYDNEY • TOKYO • WROCLAW

Editor: Charles W. Mitchell
Managing Editor: Grace E. Miller
Marketing Manager: Rebecca Himmelheber
Production Coordinator: Raymond E. Reter
Project Editor: Jennifer D. Weir
Designer: Dan Pfisterer
Illustration Planner: Lorraine Wrzosek
Cover Designer: Graphic World, Inc.,
          St. Louis, Missouri
Typesetter: Maryland Composition Co., Inc.,
          Glen Burnie, Maryland

Printer & Binder: R. R. Donnelley &
          Sons Company,
          Crawfordsville, Indiana
Digitized Illustrations: Maryland Composition
          Co., Inc.,
          Glen Burnie, Maryland

351 West Camden Street
Baltimore, Maryland 21201-2436 USA

Rose Tree Corporate Center
1400 North Providence Road
Building II, Suite 5025
Media, Pennsylvania 19063-2043 USA

Accurate indications, adverse reactions, and dosage schedules for drugs are provided in this book, but it is possible that they may change. The reader is urged to review the package information data of the manufacturers of the medications mentioned.

Printed in the United States of America

Library of Congress Cataloging-in-Publication Data

Wood, Bradford J.
     Radiology / Bradford J. Wood, Sangeeta Desai Wood.
          p.   cm.
     Includes bibliographical references and index.
     ISBN 0-683-30363-5
     1. Diagnostic imaging—Handbooks, manuals, etc.   2. Diagnosis,
Differential—Handbooks, manuals, etc.   3. Medical emergencies—
Handbooks, manuals, etc.   I. Wood, Sangeeta Desai.   II. Title.
     [DNLM: 1. Diagnostic Imaging—handbooks.   2. Emergencies—
handbooks.   3. Diagnosis, Differential—handbooks. WN 39 W873r
1998]
RC78.7.D53W66   1998
616.07′54—dc21
DNLM/DLC
for LIbrary of Congress   97-27521
                                                                                              CIP

The publishers have made every effort to trace the copyright holders for borrowed material. If they have inadvertently overlooked any, they will be pleased to make the necessary arrangements at the first opportunity.

To purchase additional copies of this book, call our customer service department at **(800) 638-0672** or fax orders to **(800) 447-8438.** For other book services, including chapter reprints and large quantity sales, ask for the Special Sales department.

Canadian customers should call **(800) 665-1148,** or fax **(800) 665-0103.** For all other calls originating outside of the United States, please call **(410) 528-4223** or fax us at **(410) 528-8550.**

Visit Williams & Wilkins on the Internet: http://www.wwilkins.com or contact our customer service department at **custserv@wwilkins.com.** Williams & Wilkins customer service representatives are available from 8:30 am to 6:00 pm, EST, Monday through Friday, for telephone access.

97   98   99   00   01
1   2   3   4   5   6   7   8   10

*To my wife, the best friend ever invented, and
to our families, who keep us in tune.
In remembrance of
Charles "Tunk" Tegtmeyer, MD,
a loud voice that many still hear.*

Preface

This book in the Williams & Wilkins House Officer Series provides medical students and junior residents with a brief overview of introductory radiologic anatomy, as well as a basic approach to organizing image analysis and decision making. The text presents a concise and minimalist approach to commonly encountered on-call and emergency medical problems. Ideas are short and simple, and everyday words replace medical jargon where possible. Emphasis is placed on emergency decision-making problems, diagnostic dilemmas, and "pearls."

Practical information is divided by systems. Differential diagnosis, helpful acronyms, and clinical pearls are presented in bullet format. Indications, clinical use of imaging modalities, and commonly asked board and attending questions are included.

Emergency, trauma, and after-hours imaging are often the most important imaging studies, with the most at stake. The beginning doctor needs a simple and digestible cookbook for approaching the cumbersome and confusing field of radiology. Exposure to a basic core of radiology with which every physician should be familiar is becoming more difficult and complex. Basic radiology provides the perfect framework for consolidating facts and placing the final pieces in the puzzle of anatomic and pathophysiologic correlation. This integration should allow for a fertile exchange of ideas among the radiologist, clinician, and student, resulting in improved patient care.

Each chapter has a similar structure, beginning with an overview of the peculiarities of that system's organized approach. Short cookbook tables outline the basics. Normal and abnormal radiographic anatomy is presented pictorially with text discussion of clinical correlates. Main principles are in **bold.** Sections close with discussions of differential diagnoses and clinical pearls. Included are helpful memorization devices, a review of basic principles, and medical trivia often asked on rounds and on boards. Only common abnormalities are included; zebras that waste time or cloud retention are omitted. Following most chapters is a brief list of preferred introductory radiology texts for more inquisitive minds.

The rapidly expanding field of radiology requires familiarity with new technology and techniques. The final chapters will briefly cover nonsystem-oriented subjects such as helical computed tomography (CT), digital imaging, imaging patients with AIDS, three-dimensional imaging, angiography with CT, magnetic resonance (MR), common patterns of neoplastic and metastatic disease, interventional radiology, and relative cost of different modalities.

Some medical schools are dissolving the required radiology rotation and replacing it with primary care rotations. Many new doctors matriculate with little or no formal radiology rotation or instruction. Up-to-date and readable student texts are needed to fill this void. This short portable text for on-call and emergency reference is preferable to wading through thick reference texts. We hope this book will encourage further study in a more complete text such as Squire and Novelline's *Fundamentals of Radiology* or *Basic Radiology,* edited by Chen, Pope, and Ott.

This book is in no way a substitute for clinical experience or for reading a more comprehensive introductory text. References have purposely been left to a bare minimum, at the risk of oversimplification or misrepresentation. Though facts, rules, and numbers are useful, they are only teaching guidelines and should not be used clinically without clinical correlation and the consultation of the attending physician and radiologist. Some of the percentages and numbers are controversial and meant to convey a range or a principle. Generalizations are only guidelines and can be dangerous if taken as absolute truths.

Defining moments in one's life are often times of great change. One of these is the first night on-call or the first days on a new clerkship. Autonomy and confidence meet head-to-head with information overload, responsibility, and fear. *Radiology* was written with this in mind as a guide through these murky waters. Remember that the joy is in the journey. Good luck!

Acknowledgments

We are grateful to the many medical students and residents at Georgetown, George Washington, and Massachusetts General who made this book possible. Thanks to the Georgetown staff: Cliff Lefteridge, Becky Zuurbier, Bolivia Davis, Craig Platenberg, Curtis Green, Cirrelda Cooper, Matthew Freedman, Bob Zeman, and Jeff Love for their dedication to teaching. Cliff, Bolivia, and Jeff also contributed films. The support and vision of our editor and friend, Charley Mitchell, were vital and much appreciated. Thanks also to our friend Peter Mueller for "feeling the fire and buying the mission."

Contributors

Shilpa Desai, MD
Resident, Ophthalmology
George Washington University Hospital
Washington, DC

Lisa Kelly, MD
Resident, Radiology
Georgetown University Hospital
Washington, DC

Michael Zalis, MD
Resident, Radiology
Georgetown University Hospital
Washington, DC

Mukesh Harisinghani, MD
Resident, Radiology
Massachusetts General Hospital
Boston, Massachusetts

Mark Ryan, MD
Fellow, Abdominal and Interventional Radiology
Massachusetts General Hospital
Boston, Massachusetts

Christopher Grady, MD
Resident, Radiology
Georgetown University Hospital
Washington, DC

Contents

   Background
   Physics/How Images Are Formed
   Communication/Language
   Organized Approach
   Modalities
   Checklist
   Relative Cost of Studies

   Chest Cookbook
   Killers: Free Air/Pneumothorax
   Technique
   Portable Chest X-Ray
   Tubes and Lines
   Pneumomediastinum/Pneumopericardium
   Pleural Effusions
   Bones and Soft Tissues
   Mediastinum/Vessels/Lymphadenopathy
   Heart
   Lungs
   Pulmonary Edema
   Pulmonary Vascular Patterns
   Alveolar Versus Interstitial Patterns
   Airspace Disease
   Other Lung Diseases
   Lung Cancer Pearls
   Pulmonary Embolus (PE)
   Segmental Anatomy

Chapter 1

Overview of Emergency and On-Call Radiology

The main goal of this handbook is to introduce imaging studies in anatomic and practical terms. The cookbook approach and the oversimplified words and pictures provide a skeleton view of a complex field, making it more digestible for the beginner. The final criteria for usefulness will be your future patients' care.

Background pathophysiology, ridiculous mnemonics, pictures with words and drawings, and, most importantly, patients themselves are the tools that help us to remember details and to better care for patients. A systematic and methodical approach is a vital foundation; learning to read an x-ray is worthless if it is not correlated with clinical information.

BACKGROUND

When evaluating an image, think anatomically, three-dimensionally, and in terms of differential diagnosis. Ask "What organs or structures lie near here? What process could tie together the clinical and imaging findings to make a single diagnosis? How might this process affect other organ systems or structures?" Imagine organs in their anatomical position because the two-dimensional images can be misleading. Remember, radiology is like real estate: location, location, location!

The organized approach to **differential diagnosis** divides the patient's history, physical, labs, and imaging into categories of diseases using the acronym, "**CIN TV**": **C** = congenital; **I** = infectious, idiopathic, inflammatory, iatrogenic; **N** = neoplastic; **T** = traumatic; and **V** = vascular.

Uncommon presentations of common problems are more likely than unusual diseases (a striped horse is probably just a striped horse and not a zebra). Always consider the possibility that something else could be mimicking your diagnosis. This way, you will learn even from your straightforward patients. Also, consider the costs and benefits of ordering a more specific test alone (such as an magnetic resonance [MR] instead of a computed tomography [CT]), if it will be done later anyway.

Familiarize yourself with normal images. Abnormalities are only recognized if you know how they differ from the normal lines and shadows. Symmetry is your friend. Look side to side for asymmetry or irregular densities (with pediatric bone films, a comparable view of the opposite limb often clarifies questions).

PHYSICS/HOW IMAGES ARE FORMED

Staring at a CT or an x-ray can be overwhelming with so many overlapping shadows. Recreate a three-dimensional image by dissecting the many pictures into small chewable parts. "**X-rays**" are images of shadows of x-rays added up along straight lines of travel. Imagine a very powerful light beam that penetrates thin or light tissue more than thick or dense tissue. The x-rays that shoot all the way through the patient (as in air or lung) make the film black (exposed). Metal and bone block more of the x-ray beam and leave a white shadow on the film (fewer x-rays make it through to the film). The many shades of gray in between give the information to decipher an image.

The edge of an object is only seen if it is next to a structure of different density (lung next to heart). This "**silhouette**" line will be sharp and well-defined if the beam is tangential to this edge, there is no patient motion, and no abnormal densities are next to this edge. Pneumonia, heart failure, pleural fluid, or any soft tissue can blur the heart or diaphragm silhouette on a chest x-ray. However, a shadow can also be from something outside of or on the patient (nipples and electrocardiogram [ECG] electrodes may look like lung nodules, and oxygen tubing or skin folds may look like pneumothoraces at first glance).

An abnormal density can be caused by anything that blocks or shadows the beam. This may occur anywhere along the beam pathway. Imagine the x-ray beam as a needle that goes through the patient. Each structure through which it passes should make you think of a different "**CIN TV**" list of possible diagnoses. Keep a short mental list of the two or three most likely.

In x-rays and CT, brightness is a measure of tissue **density** (or how many x-rays it stops, or "attenuates"). **Brightest to darkest (or most dense to least dense): metal > bone > blood > brain and soft tissue > cerebrospinal fluid (CSF), urine, and water > fat > air.**

CT numbers (region of interest or Hounsfield units) tell how much x-ray energy is stopped by a selected area on CT. Negative numbers identify fat; simple fluid (water) is between zero and 20 units; blood is characteristically about 60 units; and calcium has high numbers (Fig. 1.1). CT numbers can help diagnose cysts, benign fatty tumors, or types of kidney stones. MR images the density of water or protons in tissue.

CT is simply a fancy rotating x-ray machine where both the x-ray "gun" and the receiving "film" rotate around the patient in complete circles (or in a moving spiral in helical CT). X-rays shoot through the patient as the table moves in increments (or continuously in helical CT) through the CT donut like a high-tech car wash. The computer makes an image from each rotation (**slice**).

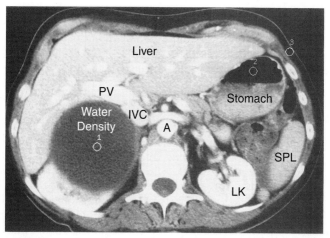

Figure 1.1. CT Numbers: water = near 0 CT#, fat = negative CT#.

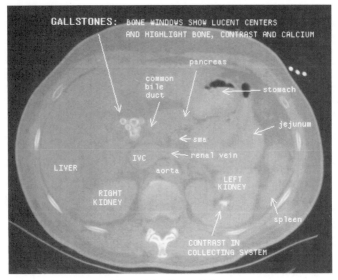

Figure 1.2. Bone windows.

We can manipulate the data to look at either bone, soft tissue, lungs, brain, or blood (**windows**) (Figure 1.2). Windows are like choosing the power of magnification in a microscope depending on what kind of tissue in which we are interested. Choosing a window "**level**" tells the computer where to assign or apply the gray scale. Choosing a window "**width**" determines how many different kinds of tissue you want to include in your gray scale, or how "wide" you want your window open. (The gray scale is the many different shades of bright and dark gray in an image.) Everything that is not in the selected window of the gray scale will be totally black or white. Thus, choosing a window is a balance between quantity and quality of information. Opening the window wide to see more kinds of tissue will result in less specific information about any particular tissue.

Window choice can be vital. CTs can be done with one or two predetermined standard windows; head CTs have bone, brain, and, sometimes, blood windows. Chest CTs have lung and mediastinal/soft tissue windows. Abdominopelvic CTs have mainly soft tissue windows. Additional windows should be examined according to the history. It may be very helpful to go to the CT scanner and play with the manual windows to look for subtle collections of blood or air which may not be well seen on the standard windows.

CT is most often used in the **emergency** or on-call setting for the head and abdomen, areas where extra windows are most helpful. Additional bone windows in the chest and abdomen can show fractures. Lung windows in the abdomen can diagnose hidden free intraperitoneal air (**free air**) from a ruptured or perforated bowel, and can also highlight hidden **pneumatosis** (air in the bowel wall) seen with bowel ischemia, infection, or infarction. The **subdural** blood in a head trauma patient may only be seen on the narrow blood windows, which your hospital may not routinely print. It may be up to you to go and ask for these.

COMMUNICATION/LANGUAGE

Films cannot be read in a vacuum. *Go with your patient to radiology, or at least make sure the reason for the exam gets conveyed somehow to the radiologist.* Too many history sheets in radiology read "rule out problem." Share the findings with the patient and include them on the chart.

Complete communication can make a big difference.

If you see an abnormality, don't jump ahead to a diagnosis. Slow your mind down and be complete! Finish the rest of your evaluation before returning to the obvious finding or you'll miss other important findings. Don't jump straight to the lung fields on a chest film. Set them aside on a shelf and look at the **less obvious things first** (foreign bodies, tubes, bones, cardiomediastinal silhouette, pulmonary vascularity, lymphadenopathy). After your analysis is complete, go back and look a second time at your **psychological "blind spots."**

Don't let the **language** of radiology confuse you. "**White**" is x-ray "density," or CT "high attenuation," or MR "high signal," or ultrasound (US) "hyperechoic," or nuclear medicine "high activity," or bright. "**Black**" is x-ray "lucency," or CT "low attenuation," or MR "low signal," or US "hypoechoic," or nuclear medicine "low activity or photopenia," or dark.

"Right" and "left" always refer to the patient's sides, not ours. **PA** (posterior to anterior) or **AP** (anterior to posterior) describe the path and direction of the x-ray beam. "Left **decubi-**

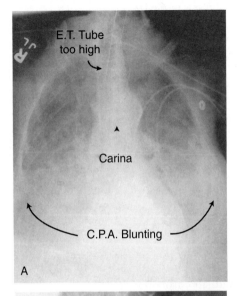

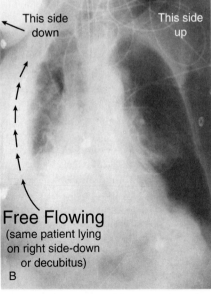

Figure 1.3. A, B. Bilateral pleural effusions.

tus'' means lying with the left side **down.** This term is commonly confused and should be replaced with "left down decubitus" when ordering studies, such as to see whether a pleural fluid collection is "**free flowing**" as opposed to "**loculated**" (separated into a pocket that does not freely communicate or shift) (Fig. 1.3). Transudates are more likely to be free-flowing, and exudates are more likely to be loculated.

Most **portable** chest x-rays are shot AP. Heart size cannot be reliably evaluated on an AP chest x-ray because the heart in the anterior chest is farther from the film and, thus, casts a larger shadow (Fig. 1.4). An unstable patient should not be transported to radiology when a portable exam might suffice. Likewise, emergent surgery should never be delayed for nonvital radiology studies.

ORGANIZED APPROACH

An organized approach starts with a quick glance to assess **technique.** Check the name, study type, position of the patient, rotation, exposure, and overall "feel." Look for correct positioning and complications of manmade tubes and lines. Then take a close-up look (inches away) and a far-back look (feet away), so you see the forest through the trees. Actually **step back** for this, and abnormalities will "jump out" off the film.

Next, perform a step-by-step review of **structures.** Lastly, go back and look at the psychological blind spots where findings are most commonly missed (on the chest x-ray, this is the lung seen "through" the heart and diaphragm, the apices, and the costophrenic angles). The clinical information is thrown in only after a methodic review of the image, so as not to skip over important surprise findings. At this point, the differential diagnosis can usually be narrowed. Ask yourself or the clinician directed questions that may lead down one or another diagnostic pathway (past medical history, medications, fever or elevated white count, history of recent travel, trauma, surgery, or radiation?).

Thinking three-dimensionally can be hard with so many two-dimensional CT or MR images. During the cookbook review, start at the top of the study and work your way down by glancing at successive contiguous images. Stack one on top of the other, and **build a three-dimensional image** in your mind. This top to bottom building should be done multiple times: once for each organ

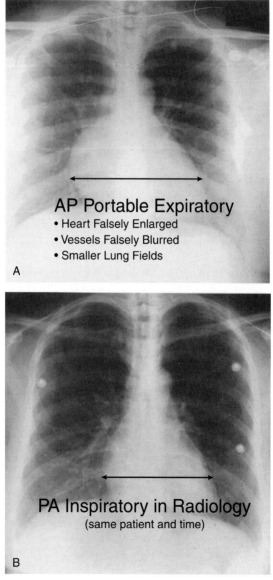

Figure 1.4. A. AP portable expiratory. B. PA inspiratory in radiology.

system. For example, in an **abdominal CT,** a series of quick glances is done while concentrating on the liver only. When this is complete, another series of glances builds a three-dimensional mental model of the spleen, gallbladder, adrenals, kidneys, pancreas, stomach, spine, and bowel. The mental reconstruction of the bowel requires changing glance directions as you follow curving loops of bowel, which go up and down. Track the loops anatomically forward and from the rectum backward. This is especially hard in the small bowel, sigmoid colon and splenic and hepatic flexures, where the direction changes may be multiple. CT and MR axial images are viewed as if you are at the patient's feet, looking up the gown, with the patient on his or her back. Thus, the patient's right is on your left and ventral is at the top of the image.

Be an alarmist; always **consider the worst** potential scenario. It is better to have considered and rejected unlikely life-threatening diagnoses than not to have thought about them at all. Think first and foremost about what can be done now at night, while you are on call, and prioritize issues. Go with your patient to radiology for emergency or on-call studies. When it's after hours or if the radiologist is on-call, radiology may be the most dangerous place in the hospital for a sick patient. **Free air** and **pneumothorax** are two life-threatening findings that can be difficult to detect, but should never be missed (Figs. 1.5 and 1.6).

Remember common **artifacts and fake-outs.** For example, two-dimensional CT images display the average density of the structures within. This can result in "volume averaging" artifacts where a less dense item is displayed as more dense because a more dense structure is averaged in as well. This can result in misdiagnosis and unnecessary procedures. Remember to **ask patients about pregnancy,** and especially limit studies in the most risky first trimester.

MODALITIES

Cine is sequential x-ray motion pictures and is used for cardiac catheterizations (and some fluoroscopy recording). **Fluoroscopy** is simply continuous x-ray television used for interventional procedures, as well as dynamic contrast studies of the GI or GU systems (upper GI, barium enema, or voiding cystourethrogram [VCUG]). Patients must be able to lie flat on a table for CT, and,

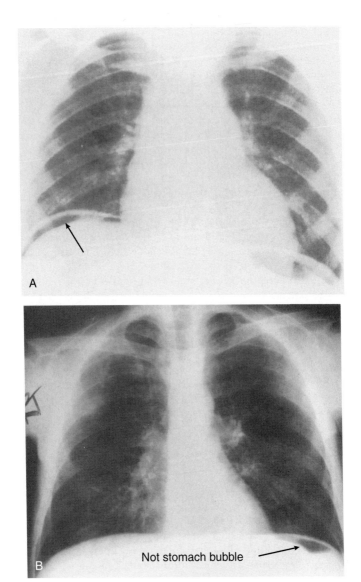

Figure 1.5. A. FREE AIR! B. FREE AIR!

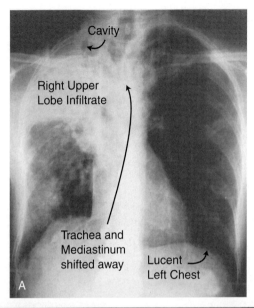

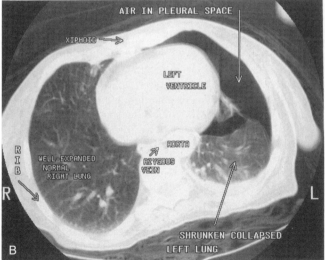

Figure 1.6.   A. Tension pneumothorax. B. Pneumothorax without tension on CT. *(continued)*

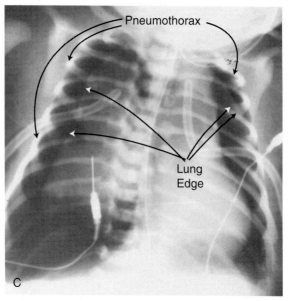

Figure 1.6. *(continued)* C. Pneumothoraces with chest tubes.

ideally, need to be able to roll around on the table for fluoroscopic barium GI studies.

Ultrasound (US) is a noninvasive quick way to look at anatomy without ionizing radiation or IV contrast. It can be done at
the bedside, in any plane or obliquity, and is exceedingly cheap
compared to CT and MR. Sound waves are sent into the body
part and an image is formed from the returning echoes. The
image depends on how long it takes an echo to return, how many
echoes are reflected from a surface (brightness), and how far the
object is from the source (position).

US may show gallstones, hydronephrosis, free fluid, ectopic
pregnancy, or an enlarged pancreas, gallbladder, or ovary. **Echo-
cardiography** is US of the heart and is performed or interpreted
by cardiologists. **Endoscopic and transesophageal US** (or transesophageal echo) may be helpful in the evaluation of thoracic
aortic abnormalities; some cardiac disorders like valve vegeta-
tions; esophageal, or pancreatic cancer staging; and pancreatic

pseudocyst or lesser sac fluid aspiration. It is usually done by gastroenterologists (or cardiologists).

MR is similar to CT, but uses non-ionizing magnetism instead of radiation. Some MR images can be viewed in any plane (**multiplanar reconstruction**) with the same resolution and clarity as axial images. This is most helpful in the brain where a lot of small structures are squashed together. Different tissue has different amounts of water (or protons) and thus different properties in a magnetic field. Proton dipoles are lined up in one direction in a magnet and then are forcefully "flipped" 90 to this direction. These "spins" are still in the magnet, and want to return back to the baseline magnet direction, or "relax." The spins "relax" at different speeds based on tissue structure and how much water is present (and whether there is any abnormal blood, protein, fluid, or edema in the area). "**Spin echo**" images (**T1, T2, and proton density**) are formed based on this principle of different "**relaxation times**" for different tissues. "**Gradient echo**" is based on varying the baseline magnet strength in different places to give more spatial information. Flowing blood is dark on spin echo and bright on gradient echo.

MR **enhancement** characteristics are similar to CT. Enhancement implies increased vascularity, or breakdown of the blood-brain barrier. IV contrast (**gadolinium**) for MR has no iodine, is not toxic to the kidneys, and can be safely used in patients **allergic to iodine.** MR first found applications with the brain, spine, and soft tissue tumors, but is becoming more and more useful in the abdomen and joints. Air, bone cortex, and old blood (hemosiderin) are black on most MR sequences. Blood in a hematoma or hemorrhage can be black or bright, depending on the age of blood products, which can be deduced from the MR characteristics. **MR angiography (MRA)** must be special ordered if there is interest in the blood vessel anatomy (narrowing, clot, vasculitis). MRA is most useful for the head and neck, aorta, and renal arteries.

Nuclear medicine ("nucs") imaging follows the administration (PO or IV) of a radionuclide (or radiopharmaceutical) and mainly assesses the physiologic function of an organ system, while giving general information on anatomy and morphology (although less specific for anatomy than other modalities). The radiopharmaceutical localizes to one or several organ systems and gives off gamma rays. The camera detects and counts these gamma rays instead of x-rays. Most studies are in two dimensions (but SPECT can look at specific slices of tissue in any plane).

The most common emergency and on-call nucs studies are for acute cholecystitis (**hepatobiliary scan**) and for pulmonary embolus ("**V/Q**" or **ventilation/perfusion lung scan**). Nucs studies also image the scrotum (testicular torsion), bowel (GI bleed), skeleton (metastases or osteomyelitis), heart (ischemic coronary artery disease), thyroid, kidneys, bladder, brain, liver/spleen, bone marrow, tumors, and abscesses.

SUMMARY

There is no substitute for looking at, talking to, and touching the patient. Many diagnoses can often be made in the absence of imaging and labs. All too often, we rely on expensive tests to confirm what we already know. Don't order tests out of curiosity or fear. **Only order tests if they will change patient management.** Think hard before you write every order.

The best teaching is often hidden in the patient's x-ray jacket, in a huge envelope of confusing studies. Take the opportunity to look at the old studies to learn the history that the patient didn't remember. **Old studies** are often the only way to tell the significance of current findings. For example, airspace disease on a chest x-ray developing over hours may be atelectasis (collapse) or pulmonary edema; over days it may be pneumonia; and over months it may be alveolar cell adenocarcinoma. They may all look similar on any one film. **A calcified lung nodule unchanged for at least 2 years is likely to be simply a benign granuloma,** whereas a growing partly-calcified nodule could be a rare "**scar carcinoma**" (an adenocarcinoma that develops within a scar or granuloma). Billions of dollars are spent wastefully on repeated studies, studies already done at another hospital, or studies done that can't reasonably change the patient's management or awareness of the prognosis.

Once again, stick to your **routine.** Use a mental or written **checklist** which includes a) using **quick glance and technique;** b) methodically reviewing **structures;** c) ruling out **life-threatening** processes (free-air and pneumothoraces); d) rechecking "**blind spots**"; e) using "**CIN TV**" to put the puzzle together; f) checking for **old films;** and g) asking questions. Don't be shy, talk to the patient, radiologist, radiology technologist, and the patient care team. This is the fun part and why we are all here.

CHECKLIST

Quick glance	Name, technique
Structures	Anatomy, tubes, calcifications
Fast killers	Free air, pneumothorax
Slow killers	Abscess, infection, tumor
Blind spots	Check twice
Differential diagnosis	"CIN TV"
Old films	

RELATIVE COST OF STUDIES

Pulmonary angiogram	$$$$$$$$$$$$$$
CT abdomen / pelvis	$$$$$
CT head	$$$$$
Needle biopsy	$$$$$$$$ +
Barium enema / upper GI	$$$$$$$$ +
Abdomen x-rays	$ +
Chest x-rays	$ +
C-Spine x-rays	$$
V/Q scan	$$$$$$$$ +
Bone scan	$$$$$ +
IVP	$$$
MR brain, abdomen, spine, or shoulder	$$$$$$$$$$$$$$$$$$$$$
Mammogram	$$

(Mammograms are the single most important preventative tool of radiology: Every year or 2 after age 40, every year after age 50, or younger if there is a family history.)

Suggested Readings

Chen MYM, Pope TL, Ott DJ, eds. Basic radiology. New York: McGraw-Hill, 1996.

Daffner RH. Clinical radiology: the essentials. Baltimore: Williams & Wilkins, 1993.

Dahnert W. Radiology review manual, 2nd ed. Baltimore: Williams & Wilkins, 1993.

Felson B. Chest roentgenology. Philadelphia: WB Saunders, 1973.

Meschan I, Ott DJ. Introduction to diagnostic imaging. Philadelphia: WB Saunders, 1984.

Ravin CE, Cooper C, Leder RA, eds. Review of radiology, 2nd ed. Philadel-
phia: WB Saunders, 1994.
Squire LF, Novelline RA. Fundamentals of radiology, 4th ed. Cambridge:
Harvard University Press, 1988.
Troupin RH. Diagnostic radiology in clinical medicine, 2nd ed. Chicago:
Mosby Year Book, 1980.
Weist P, Roth P. Fundamentals of emergency radiology. Philadelphia:
WB Saunders, 1996.

Chapter 2

Chest

The checklist approach can be applied to chest x-rays, which are the single study with which you need to be most familiar. Your proficiency at reading a chest film will directly impact your patient. Rule number one: don't miss the quick killers; free air or pneumothoraces. Rule number two: don't miss the slow killers; early pneumonias or cancer (when they can potentially be treated). Use your mental checklist. Dissect the film piece by piece with your eyes. Look at only one thing at a time. Don't let your eyes and mind race (Figs. 2.1–2.6).

CHEST COOKBOOK

Killers	Free air & pneumothorax
Technique	Portable, rotation, & inflation
Man-made	Tubes & lines
Mediastinum, hila, aorta, & heart	Adenopathy & cardiomegaly, aortic dissection, & aneurysm
Costophrenic angles	Pleural effusion
Bones & soft tissues	Fractures, mets, & stones
Lungs	Infiltrates, densities, & nodules

KILLERS: FREE AIR/PNEUMOTHORAX

First, do the **QUICK GLANCE** for life-threatening **free air** under the diaphragm or in the pleural space (**pneumothorax**). This is your chance to be a real star, but missing pneumothoraces or free air can haunt you. Free air may indicate a hole in the bowel (perforation) that might need emergent surgery. The stomach bubble may look like free air under the diaphragm (see Fig. 1.5). If there is doubt, a decubitus film may clarify. A big or symptomatic pneumothorax needs a chest tube to reexpand the collapsed lung (see Fig. 9.1). Pneumothoraces and free air can be especially

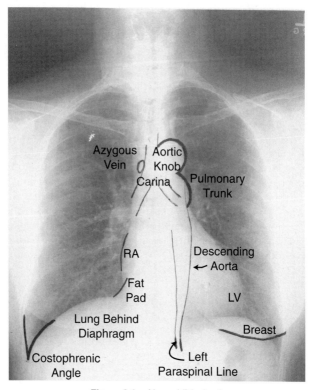

Figure 2.1. Normal PA chest.

subtle. Look at these again and again, and then some more. Portable exams are not as sensitive, but are better than nothing in the ICU, bedridden, or unstable patient.

A **pneumothorax** has a peripheral dark crescent with no lung marks peripheral to a fine thin line (Figs. 1.6 and 6.10). These are commonly due to chronic obstructive pulmonary disease (COPD), other lung disease, central venous line attempt, thoracentesis, rib fractures, trauma, positive pressure ventilation, surgery, or they can be spontaneous. Primary spontaneous pneumothorax is more common in tall, thin, young adult smokers, and results from popping a bleb. A **skin fold** simulates a pneumotho-

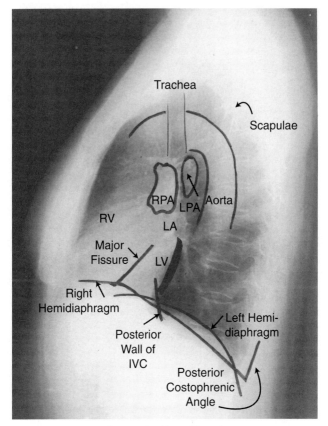

Figure 2.2. Normal lateral.

rax, but has lung markings peripheral to the line. The skin fold line is ill-defined on one side and slightly thicker than a pneumothorax because it is actually a ridge and not a line at all (Fig. 2.7). Skin folds also may extend beyond the pleural space.

If you see a pneumothorax, always look for an element of "tension," where the heart, mediastinum, or diaphragm shifts away from the increased pressure from the pleural air (see Figs. 1.6 and 2.8). A **tension pneumothorax** is an emergency because

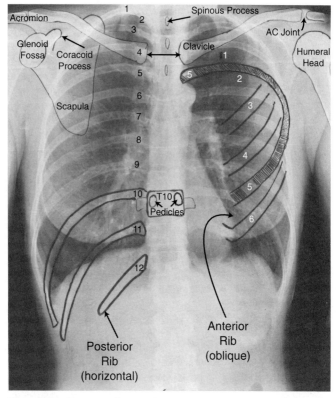

Figure 2.3. Chest x-ray bones. Note slight rotation (clavicle to spinous process distance).

the lung completely collapses and the pressure squashes the vena cavae, impairing venous return to the heart, and decreases cardiac output. A chest tube solves the problem, but a needle in the second intercostal space in the mid-clavicular line may be lifesaving and quicker. Recurrent pneumothoraces can be sclerosed (visceral and parietal pleura glued together with fibrosis) with tetracycline, bleomycin, or talc.

• A horizontal line with a pneumothorax indicates the air-fluid level of a hydropneumothorax (fluid + air) or, less likely, abscess or bronchopleural fistula.

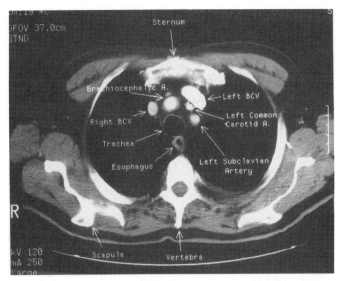

Figure 2.4.

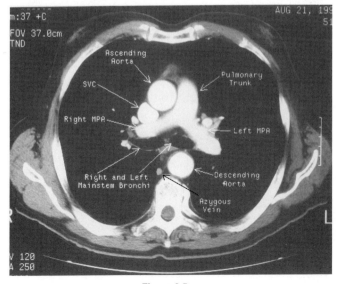

Figure 2.5.

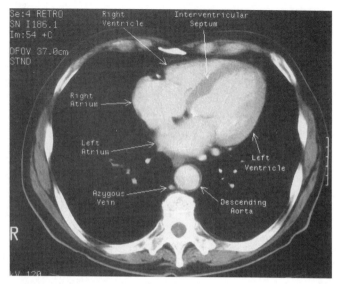

Figure 2.6.

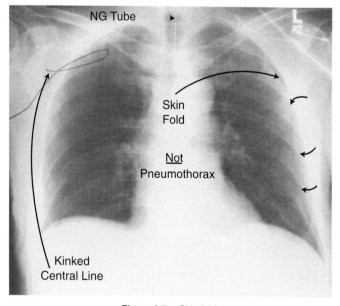

Figure 2.7. Skin fold.

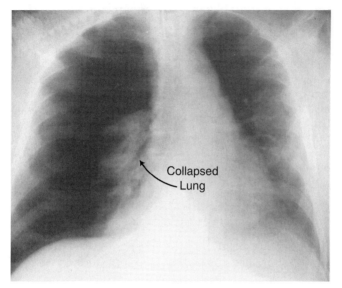

Collapsed Lung

Figure 2.8.   Tension pneumothorax. Mediastinum shifted to left.

• Catamenial pneumothorax is related to menses and is associated with pleural endometrial implants.

TECHNIQUE

Make sure the **NAME** plate matches the film. Check **TECH-NIQUE.** Is it a posterior to anterior (PA) or a portable anterior to posterior (AP) film? Is the film too bright or dark? (As a test for **exposure,** you should barely see the faint disc spaces behind the heart.) Apparent infiltrates and interstitial disease may come from a light (underexposed) or grainy (too much contrast) film. Infiltrates may be missed behind the heart and diaphragm with underexposure. Are the lungs well-**inflated**? (Count 8 to 10 posterior ribs.) A shallow breath or expiration film (< 8 posterior ribs) results in the lung vessels and markings squashing together or crowding, which can simulate airspace disease or edema. As a test for **rotation,** make sure the medial clavicles are near the midline spinous processes (Fig. 2.3). Rotation can cause false cardiomediastinal enlargement. Make sure the lung apices and costophrenic angles are fully included on the film.

PORTABLE CHEST X-RAY

The **portable** AP chest x-ray provides incomplete information. Thus, a complete PA and lateral are done in radiology and should be ordered whenever possible. Transporting the acutely ill or unstable ICU or ER patient, however, may not be safe or possible. Many normal portable films will have the false appearance of mild heart failure, or mild interstitial disease, because of underinflation, magnification, and blurring from scatter. Portable films will be less sensitive for free air, pneumothoraces, effusions, infiltrates, and nodules. The heart and mediastinum cast bigger shadows, and the vascular structures and interstitial lines are more prominent (Fig. 1.4).

- CT is much more sensitive than is chest x-ray in the mediastinum for lymphadenopathy, nodules, calcium, hilar masses, and blood.
- Hang films as if the patient is facing you.
- A leaning backward (lordotic) chest x-ray moves the overlying clavicles up, so the apices can be better seen.
- An expiratory film is most sensitive for pneumothorax.
- A normal expiratory film may falsely look like congestive heart failure (CHF) (see Fig. 1.4).
- A decubitus film may clarify the suspicion of effusion, free intraperitoneal air, or pneumothorax, especially in the intensive care unit (ICU) or infant patient.
- An inactive patient may have unilateral or asymmetric CHF from lying on one side.
- CHF in the setting of underlying COPD or chronic lung disease may result in strange patterns of asymmetric or focal edema.
- It's easier to count ribs close to the spine (the posterior aspects of the ribs are more horizontal) (Fig. 2.3).

TUBES AND LINES

Next, check for **MAN-MADE** things like foreign objects and appropriately positioned life-support devices (Fig. 2.9). Are tubes *on* or *in* the patient? Endotracheal (**ET**) **tube** tips should be between 2 and 6 cm above the carina, partly because neck flexion or extension can move ET tubes by 2 cm each way. To decompress the stomach and prevent aspiration, the nasogastric (**NG**) **tube** tip and side-port should be well below the diaphragm (and, thus,

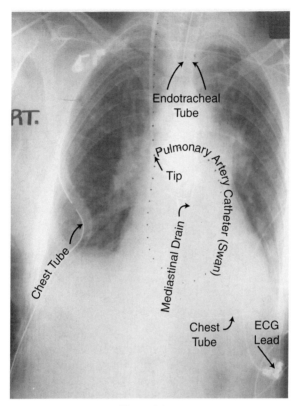

Figure 2.9.

below the gastroesophageal junction). This is true for suction or for giving medication by NG. If it is a feeding tube (some have metallic tips), it should be at least in the duodenum, to the right of midline. Keeping the patient right side down may assist advancement into the duodenum.

Central venous catheter tips should reside in the superior vena cava (SVC) or right atrium. If they tickle the right ventricle, ventricular ectopy or tachycardia may result; in the right atrium, supraventricular arrhythmias may result rarely. If they are not far enough in (tip in subclavian or jugular), vein valves may prevent blood return. Active cardiac pacemaker wires should be without

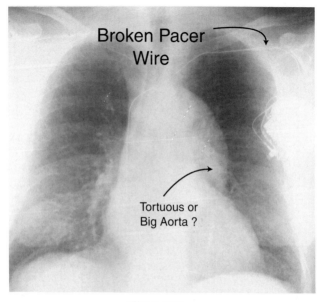

Broken Pacer
Wire

Tortuous or
Big Aorta ?

Figure 2.10.

disruption (Fig. 2.10). **Chest tube** tips and side-ports belong in the pleural space where they can drain fluid, pus, or air collections. **Swan-Ganz catheter** tips should be in a central pulmonary artery. Peripheral placement more than an inch beyond the large central vessels may block a vessel, causing a lung infarct. If it is not far enough out, it won't "wedge" and pressure tracings will be wrong. A lot of the confusing metal wires in ICU patients actually are outside the patient. Stray pieces of catheters or wires may embolize to the lungs or be a nidus for infection or clot, so hold onto wires when using the Seldinger technique!

Chest films are usually ordered following major tube manipulations, thoracentesis, NG tube placement, ET tube intubation, or attempts at central venous catheter (central line) placement. A slightly misplaced subclavian vein access attempt may cause a pneumothorax or a hematoma that widens the mediastinum. An NG tube in the airway (usually right bronchus) may cause aspiration (Fig. 2.11), so always look at the film before using the tube! (This is especially important in patients with impaired mental

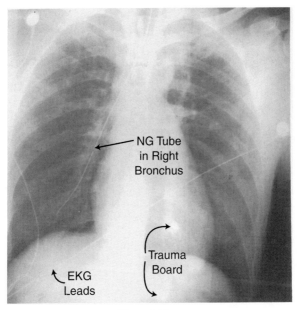

NG Tube
in Right
Bronchus

Trauma
Board

EKG
Leads

Figure 2.11.

status.) Remember that tubes and lines may be a source of fever or infection. In the ICU, this might be from sinusitis in a patient with an NG tube, or from a central line that has been in too long.

ET tube tips placed too far down usually selectively ventilate the right lung, causing a white left lung from collapse or atelectasis (Fig. 2.12). The tube tends to go to the right because the right mainstem bronchus is bigger than the left, and takes off at less of an angle. This is also why aspiration tends to occur on the right (posterior basal segment of the right lower lobe in an upright patient, and posterior segment of the right upper lobe in a supine patient). **Aspiration pneumonitis** occurs initially from the chemicals and acidity of the stomach juice, and later may become infected. The anatomy also explains why kids tend to **aspirate foreign objects** into the right lung.

The right bronchus (right upper lobe bronchus) takes off higher than the left and is thus called the ''eparterial'' bronchus (epi-arterial = above the artery). Likewise, the left bronchus (or

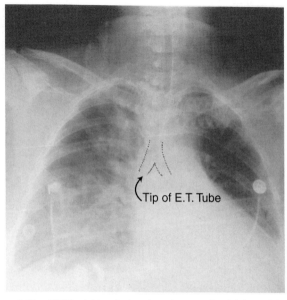

Figure 2.12. ETT in right mainstem bronchus causing left lower lobe collapse (blurred aorta and left hemidiaphragm).

left upper lobe bronchus) is the "hyparterial" (hypo-arterial = below the artery) bronchus. Now back to our checklist.

PNEUMOMEDIASTINUM/PNEUMOPERICARDIUM

Look for **air** (a dark line-lucency) in the pericardium or mediastinum (Fig. 2.13). (You can sometimes tell these apart because pneumopericardium stops at the level of the great vessel origin, whereas pneumomediastinum may continue above this. Also, pneumopericardium will change sides with position.) **Causes** include pressure ventilation, trauma, esophageal perforation, and infection.

PLEURAL EFFUSIONS

The **pleura** is an empty sac. Abnormal fluid first collects in the more dependent sac of the posterior **costophrenic angle** (CPA),

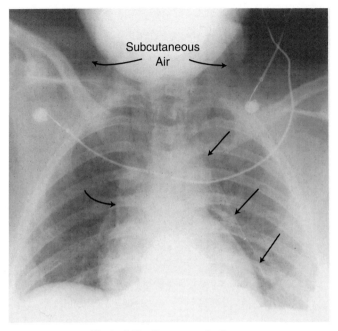

Figure 2.13.   Pneumomediastinum.

followed by the lateral CPA's sac. The posterior sulcus is lower
than the anterior. (This requires you to look behind, through,
and below the diaphragm on the frontal film to see the lower lobe
lung tissue) (Fig. 2.14). Pleural effusions blunt the CPA and may
have a meniscus (density curving up) at the edge of the effusion
where it sits up against the pleura. As little as 25 mL of fluid may
be seen on the lateral film, but it may take up to 200 to 300 mL
to see blunting of the lateral CPA on a PA film. **Decubitus films**
can tell if an effusion is ''free flowing'' versus ''loculated'' and
thus immobile (see Fig. 1.3). They also may help you determine
if an effusion is big enough to tap (thoracentesis). Apparent ele-
vation of the diaphragm on one side may be from phrenic nerve
paralysis or from a subpulmonic (below the lung) effusion. An
effusion may change shape or ''layer'' out along the lowest part
of the chest on a decubitus film (see Fig. 1.3). Old films or decubitus films may differentiate pleural effusion from chronic pleural

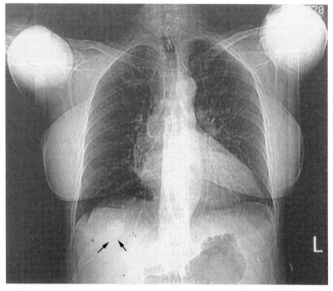

Figure 2.14. Nodule below diaphragm in right lower lobe.

fibrosis or scarring, which usually has much less clinical significance.

- An abscess is in lung parenchyma, whereas an **empyema** is a loculated, infected pleural collection, often requiring chest tube drainage.
- Ultrasound can confirm small or loculated pleural effusions and guide needle placement for thoracentesis or chest tube placement for empyema.
- Pleural-based abnormalities form obtuse angles (>90) with the lung-pleura edge, whereas lung-based masses form acute angles (<90) with the edge.
- The right cardiophrenic angle (between the heart and diaphragm) has many strange, but normal appearances (common mistake).
- The minor fissure is the floor of the right upper lobe and the roof of the right middle lobe (Fig. 2.2).
- A "**pseudotumor**" looks like a mass, but is just fluid in the fissure. (The clue is the teardrop or lens shape with borders tapering within the fissure, often best seen on the lateral film.)

- Meigs' syndrome is a benign ovarian mass with pleural effusion and ascites.
- Mesothelioma is a pleural tumor related to asbestos exposure.
- Left lower lobe atelectasis is normal for a postoperative coronary artery bypass graft (CABG) patient.

BONES AND SOFT TISSUES

Bones and soft tissues are easy to forget. Are there rib fractures or bone metastases? How about squashed vertebral bodies (compression fractures), calcifications of chronic pancreatitis, calcified gallstones, kidney stones, or bone lesions? Look at the edge of the film. If the lungs are clear, don't get happy and miss the rib or spine metastasis. Don't just stare; ask yourself questions.

The ribs are easier to analyze quickly if you don't trace each rib around; rather, look at sections of all the ribs as a group (Fig. 2.3). Look at all the posterior ribs first, then all the anterior ribs, and finally the lateral aspects, which is where most fractures occur. When ribs fracture, pneumothorax, hemothorax, lung contusion, and pathologic bone should be excluded (Fig. 2.15).

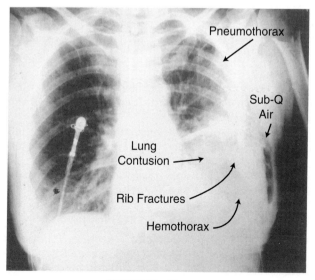

Figure 2.15. Trauma.

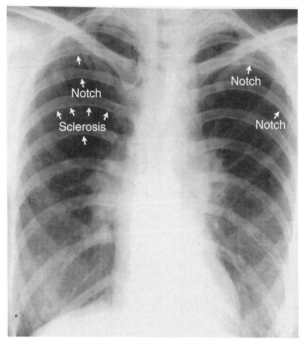

Figure 2.16. Aortic coarctation with rib notching and sclerosis from collateral intercostal vessels.

- Ribbon ribs (thin) and notching along the bottom of ribs are seen with **coarctation** of the aorta from big collateral intercostal vessels (Fig. 2.16).
- A mastectomy may make one lung look darker than the other.
- A cervical rib (from C7) may compress a vessel, causing upper extremity pain.
- A fracture in the first two ribs is associated with subclavian vessel injury and may need an arteriogram to evaluate.

MEDIASTINUM/VESSELS/LYMPHADENOPATHY

Make sure the basic structures are normal in size and contour without extra lumps or bumps. When visible, the **azygous** vein should be less than 10 mm on an upright PA chest x-ray (Fig. 2.1). Greater than this may mean volume overload, tricuspid regurgitation, right heart failure (cor pulmonale), or anything elevating venous pressure. The azygous vein sits just to the right of the right mainstem bronchus origin, and its enlargement is the radiographic correlate of jugular venous distention.

The **trachea** is a dark tube near midline. A substernal goiter commonly shifts the trachea to the left, and traumatic aortic rupture or leak can shift it to the right. Aortic rupture or leak also causes an apical pleural density (apical cap) and widening of the superior mediastinum (Fig. 2.17). The **aortic arch** (knob) is irregular or big in **dissection, aneurysm, or traumatic rupture or leak** (Fig. 2.18). It is small, inapparent, or indented in coarctation. If there is any question of mediastinal widening, blood, or aortic injury, a normal chest CT can preclude the more invasive aortogram. If there is no mediastinal "dirty fat" around the aorta on contrast CT, an aortic injury or mediastinal bleed is unlikely. Fat is usually dark on CT, but **dirty fat** is streaked with brighter

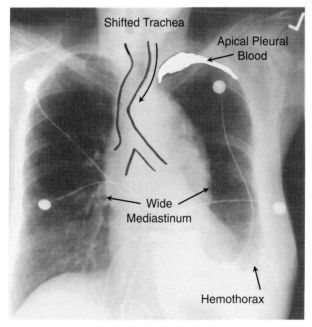

Figure 2.17.   Aortic rupture.

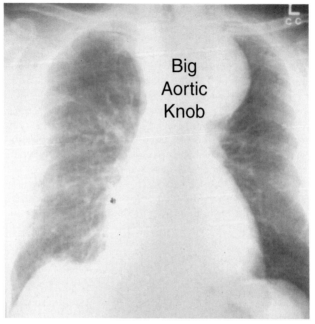

Big
Aortic
Knob

Figure 2.18.   Dissection, aneurysm, or rupture?

inflammatory cells or blood. If there is mediastinal blood, an
aortogram is the next step.

Big nodes (**lymphadenopathy**) may be infectious, inflammatory, or neoplastic. Big subcarinal nodes can splay the carina
wider (so will a big left atrium). The dense stripe (right paratracheal line) immediately to the right of the trachea should be less
than 4 mm wide. Any bigger suggests adenopathy (right paratracheal and bilateral hilar adenopathy is a hallmark of **Sarcoid,**
usually in middle-aged black females) (Fig. 2.19).

Hilar adenopathy is best seen on the lateral film as an enlarged, lumpy-bumpy, lobulated hilum (Fig. 2.20). Hilar masses
may be solitary and smoother. The "**aorticopulmonary window**"
between the aortic knob and the pulmonary trunk is normally
empty and forms a sharp angle; however, the window fills in with
big, lumpy aorticopulmonary nodes. A big lumpy azygous vein
may in fact be azygous adenopathy. **Bilateral hilar adenopathy**

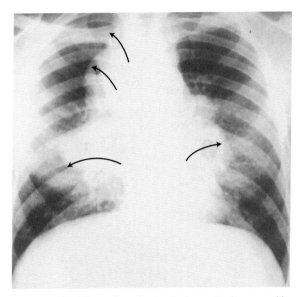

Figure 2.19. Hilar and paratracheal adenopathy from sarcoid.

suggests lymphoma, sarcoidosis, or tuberculosis, whereas unilateral suggests primary lung cancer.

There are four vertical **mediastinal** lines: two faint paraspinal lines (immediately next to the vertebral bodies), the right lung posterior-medial edge (azygoesophageal recess), and the descending aorta on the left (Figs. 2.1 and 2.3). Any deviation from straight and well defined (like focal bulge or fuzzy) can signify pathology. A tortuous descending aorta can be seen with old age, hypertension, and aortic regurgitation. Posterior mediastinal masses can cause paraspinal bulge. These are very faint, subtle lines and will take practice to see. They may not be seen at all on portable films.

The **mediastinum** is divided into posterior, middle, anterior (and superior) parts, divided by the heart, trachea, and spine. **Middle mediastinal** masses may come from the esophagus or tracheobronchial tree (congenital cysts). **Posterior mediastinal** masses are usually neural (neuroma, neuroblastoma, or neurofibroma) or congenital. Less common posterior masses include

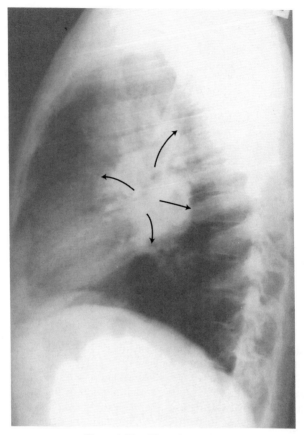

Figure 2.20. Hilar adenopathy.

esophageal tumor, hiatal hernia, discitis with abscess, extramedullary hematopoiesis, and hematoma from spine fracture. Always consider **adenopathy, lymphoma, and aneurysm** in the differential diagnosis (DDX) of *any* **mediastinal mass** in any location.

• DDX **anterior mediastinal** mass = "**Four Ts**"—thymoma, thyroid mass, teratoma, and (terrible) lymphoma (Fig. 2.21).
• Hodgkin's staging is based on bilaterality, whereas non-Hodgkin's lymphoma staging is based on crossing the diaphragm.

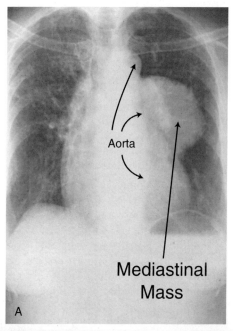

Aorta

Mediastinal
Mass

A

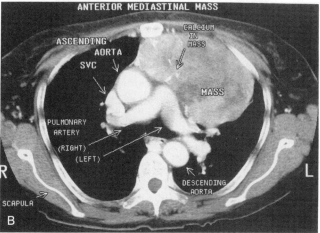

B

Figure 2.21.

• Mediastinoscopy may sample paratracheal, right tracheobronchial, or anterior subcarinal nodes.
• 15% of myasthenia gravis patients have thymomas, and 50% of patients with thymomas have myasthenia.
• In a child, a triangular anterior mediastinal mass (thymic sail) that is soft enough to be scalloped or indented by the ribs (thymic wave) is a thymus.
• MR, CT, or angiography can diagnose the aortic dissection, aneurysm, or rupture that causes a big aortic arch on the chest x-ray.

HEART

The adult **heart** shadow should be less than ½ the thoracic diameter (the **cardiothoracic ratio,** measured rib to rib on an upright PA film) (see Fig. 1.4). Greater than this indicates **cardiomegaly or pericardial effusion.** However, this is not a reliable measurement on AP, supine, or underinflated films or with obese patients (all of which falsely raise the apparent size of the heart and mediastinal shadows). The normal adult heart is shaped like a baby boot. A pericardial sac full of amorphous fluid (**pericardial effusion**) will be big and Hershey kiss shaped (Fig. 2.22). A pericardial effusion may show an "Oreo cookie" sign on a lateral film: the dark epicardial and substernal fat stripes (chocolate) are displaced from each other by a dense stripe of pericardial fluid (vanilla cream).

A big left ventricle (LV) moves the apex to the left on the PA film and bulges towards the spine on the lateral film. A big LV may indicate ischemic or dilated cardiomyopathy, aortic stenosis, HIV cardiomyopathy, or even the high-output state of sickle-cell anemia. A big right ventricle on the lateral film will fill in the normally dark lung density area behind the sternum (retrosternal clear space), but is difficult or impossible to differentiate from LV enlargement on the PA film. Both ventricles are often enlarged anyway.

A big right atrium may bulge the right heart border. Two edge shadows over the *right* heart (**double density**) may indicate *left* atrial enlargement. Three bumps along the upper left heart border (aortic knob, big pulmonary trunk and big left atrial appendage) with pulmonary edema and a big heart equals **rheumatic mitral valve disease** (stenosis + / − regurgitation) (Fig. 2.22).

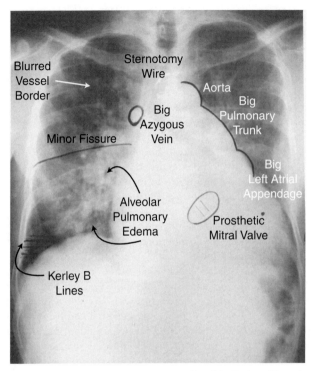

Figure 2.22. CHF with cardiomegaly ± pericardial effusion. (From rheumatic mitral valve disease.)

The pulmonary trunk is contiguous with the left heart border and becomes flat or concave by 35 years of age. A bulging trunk after 35 years age is abnormal. The aorta and innominate/subclavian vessels may become more bent (tortuous), bigger (ectatic), and calcific with age. This may cause apparent widening of the superior mediastinum.

LUNGS

When the lung is diseased or attacked, it responds in only five ways: swelling (edema, CHF, ARDS), secreting (blood, pus, water, cells), bleeding, collapsing (atelectasis), or losing parenchyma (emphysema, scarring).

PULMONARY EDEMA

Now examine the **pulmonary vascularity** for early **pulmonary edema.** In the acute setting, the PA upright chest x-ray vascularity correlates fairly closely with degree of heart failure (and with "pulmonary capillary wedge pressures" [PCWP] from the pulmonary arterial catheters [Swan-Ganz]). Normal wedge pressures are less than 12 mm Hg. Early LV failure (from a myocardial infarct, congestive heart failure, cardiomyopathy, etc.) results in thick, swollen upper lobe vessels and narrowed lower lobe vessels, a phenomenon called "**redistribution**" (or cephalization). If the peripheral upper lobe vessels are bigger than your pencil (on a conventional film), and equal to or bigger than the lower lobe vessels, then redistribution is likely present. Redistribution *cannot* be determined from AP, portable, supine, or poorly inflated films, where mild edema cannot be detected. Luckily, the worse stages of pulmonary edema (or "pulmonary venous hypertension") *can* be seen on the AP, portable, and supine films.

Interstitial disease is the next worse stage of **pulmonary edema,** manifest by increased lines and markings with blurred vessel edges (you cannot draw a line with your pen around these indistinct vessel edges). The same interstitial and lymphatic engorgement results in thickened bronchial walls (**peribronchial cuffing**—seen on end as circles with borders thicker than your pen can draw), thin lines at the lung bases extending to the lung edge (swollen lymphatics called "**Kerley B lines**"), and fluid in the fissures and pleural space (right effusion more common than left) (Fig. 2.22).

The next and worst stage of pulmonary edema is diffuse fluffy **alveolar (or airspace) disease,** often worse in the perihilar region. Interstitial edema eventually fills the interstitium and overflows into the **alveoli** (eventually causing complete lung white out). Often, superimposed pneumonia can be impossible to exclude and the patient may need antibiotics until it is clarified. Sorting it out may require a comparison with old films or history to assess chronicity, a trial of diuresis to treat the edema, checking for fever or white cell count, or simply time passage. **Edema and atelectasis usually come and go quicker (hours) than pneumonia (days),** which may be the only important clue. Different people and different institutions have different meanings for the words "**infiltrate**" and "**airspace disease.**" Ask what is implied or meant by these ambiguous terms.

Stage of CHF	Description	PCWP (<12 is normal)
Redistribution	Big upper vessels	Teens
Interstitial	Blurred vessels	20s
Alveolar	Fluffy	30s

PULMONARY VASCULAR PATTERNS

Other more subtle patterns of vascularity are less common, but may be helpful. The "**pruned tree**" appearance (big central and hilar vessels with rapidly tapering small peripheral vessels) is seen with pulmonary arterial hypertension from any cause including chronic lung disease, COPD, right heart failure (cor pulmonale), multiple pulmonary emboli (PE), connective tissue disease, or even Eisenmenger's syndrome (shunt reversal).

The "**overcirculation**" pattern has big, well-defined vessels centrally and peripherally, possibly indicating a septal defect in the heart (causing a left-to-right heart shunt) or a patent ductus arteriosus. This is most commonly an atrial septal defect in the adult population.

ALVEOLAR VERSUS INTERSTITIAL PATTERNS

Finally, pull the lungs down from the shelf and look for infiltrates, collapse, nodules (<2 cm), masses (> 2 cm), or lung disease. Lung disease and densities on x-rays are often divided into "interstitial" and "alveolar" (or airspace disease). **Airspace disease is further divided into infiltrate** (implying a filling process like edema or infection) **and** collapse (**atelectasis**), which may look identical or coexist. **Airspace disease** is soft, patchy, cloud-like, ill-defined, or confluent density that tends to conform to lung segments (Fig. 2.23); whereas **interstitial disease** is stringy, linear, granular, honeycomb, reticulated, or non-confluent web-like interlaced lines (see Fig. 9.2). This is a very important, subtle, and difficult differentiation; however, these two often coexist or can't be differentiated (Fig. 2.24), so don't get bogged down in this.

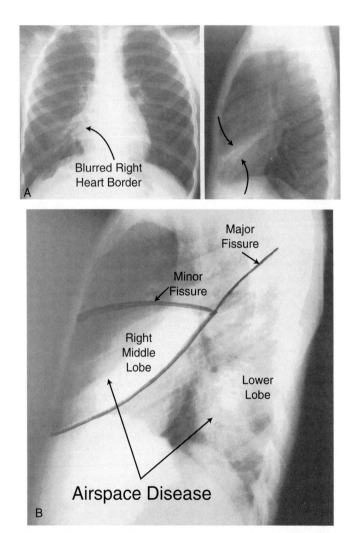

Figure 2.23. A. Right middle lobe pneumonia or atelectasis. B. Normally, spine gets darker as you go down on the lateral. (Lighter = airspace disease.)

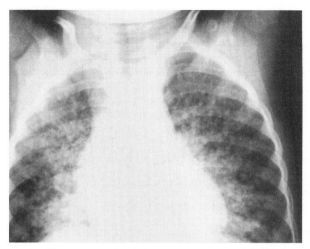

Figure 2.24. Mixed alveolar interstitial pneumonia.

Do know the basic difference between these two patterns and their common causes.

Pattern	Description	Cause:
Alveolar	Fluffy, cloud-like, segmental	Blood, pus, water, cells
Interstitial	Stringy, honeycomb	Fibrosis, lymphangitic tumor, mild CHF, viral pneumonia

- **DDX of alveolar disease: blood, pus, water, cells.**
- **DDX of interstitial disease:** mild heart failure, chronic lung disease (interstitial fibrosis), lymphangitic tumor spread, viral or interstitial pneumonia, collagen-vascular disease, pneumoconioses or inhalational, silicosis, radiation, sarcoidosis, drug reaction (the list is too long to remember).
- "Lymphangitic tumor spread" is most commonly from lung, breast and stomach or GI cancers.
- **Infection, pulmonary edema, or chronic lung diseases can all cause interstitial, alveolar, or mixed disease.**

Remember that **airspace disease** next to soft tissue will blur that tissue border that is normally well seen (when outlined by normal lung air). A profile is only seen when two tissues of *different* density are next to each other. This blurring principle is called the "**silhouette sign.**" The right heart border is blurred by adjacent right middle lobe airspace disease. This principle also applies to hilar masses that tend to blur the hilar structures, whereas masses anterior or posterior to the hilum do not. A dark tube within an alveolar infiltrate is an "**air bronchogram,**" where an air-filled bronchiole is surrounded by dense airspace disease (Fig. 2.25).

AIRSPACE DISEASE

Atelectasis

Airspace disease that is from **atelectasis** will create a **vacuum** and pull mobile things towards the area of collapse. These include the heart, mediastinum, hila, diaphragm, fissures, and lung markings. Lung marks are made of vessels, bronchi, lymphatics, and interstitium. Keep in mind that collapse may coexist with infiltrate or effusion. A "**post-obstructive**" collapse occurs distal to a hilar or bronchial mass that compresses the bronchus. If this is infected, antibiotics may be needed to cover anaerobic infections also (postobstructive pneumonia).

For example, **right lower lobe** volume loss pulls the right hilum down, pulls the hemidiaphragm up, and pulls the major fissure backwards on the lateral film. The diaphragm blurs, but the right heart border (right atrium) remains sharp, well seen through the density of the airspace disease (because it is the right middle lobe that touches the heart, in the front of the chest). Thus, it is **right middle lobe airspace disease** that **blurs the right heart border** (Fig. 2.23). On the lateral, the right middle lobe and the lingula (of the left upper lobe) project over the heart. **Left lower lobe** airspace disease will blur the descending aorta and the medial left hemidiaphragm, and may be seen behind, or *through* the heart (retrocardiac).

On the normal lateral film, the spine should become darker lower down the thorax (Fig. 2.23). In the presence of lower lobe airspace disease, the lower thoracic spine may be equal to or brighter than the levels above. (The lower lobes are triangular and more posterior).

The lobes tend to collapse toward the mediastinum. **Left upper lobe** collapse is most difficult to figure out; it results in

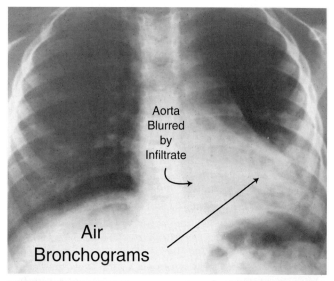

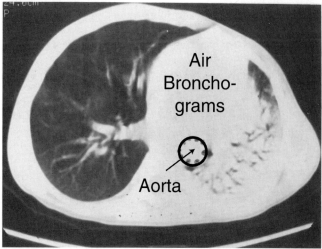

Figure 2.25. Air bronchograms.

increased background density in the left upper chest with a blurred left heart border and a retrosternal band of increased density on the lateral film, outlined by the major fissure bending to the front.

- Atelectasis can be airspace disease; volume loss; or, simply, flat, discoid bands.
- The left hilum is higher than the right.
- Mucous plugs can cause focal atelectasis, especially in cystic fibrosis, ventilated patients, and infants.
- Foreign body in the airway can cause air trapping with hyperinflation and asymmetry via a ball-valve effect, or can cause collapse.

Infiltrate

Airspace disease from **infiltrate** causes white alveoli that fill with one of four things: **blood, pus, water, or cells.** "Blood" may be from contusion, anticoagulants, thrombocytopenia, Goodpasture's, infarct (PE), or a bleeding problem. "Pus" or pneumonia may be from bacteria; virus; aspiration; or, less likely, fungi, mycobacterium, or even parasites. "Water" is pulmonary edema, which usually is diffusely distributed, sometimes in a "batwing" or perihilar pattern. "Cells" implies cancer. (Alveolar cell carcinoma is a subtype of adenocarcinoma and may present as chronic airspace disease).

Water and pus are the most common. Although often cardiogenic, remember that pulmonary edema can also be neurogenic, uremic, septic, iatrogenic (fluid overload), or even from inhalation injury (smoke or toxin), near-drowning, heroin overdose, altitude sickness, radiation, aspiration, transfusion, or drug reaction.

Infiltrate and Consolidation (Pneumonia)

Consolidation is a very dense, airless part of lung that is filled with pneumonia. Community-acquired lobar pneumonia is classically from pneumococcus (strept). Bacterial pneumonias tend to cause dense **airspace** disease with volume loss in a lobar or segmental distribution, organizing according to lobar boundaries (fissures). **Bacterial** pneumonias may also spread via the bronchi, causing "bronchopneumonia" which involves the alveoli of multiple bronchi in a patchy, multifocal, scattered distribution. **Viral** pneumonias, on the other hand, classically cause bilateral **interstitial** perihilar infiltrates with peribronchial cuffing and hyperinflation. Viral pneumonias may also cause mixed airspace and interstitial infiltrates (like *Mycoplasma* typically does). **Fungi** tend to be nodular, but have many different appearances.

Hospital-acquired **pneumonias** are usually gram-negative and necrotizing (needing broad-spectrum antibiotic coverage). Supposedly, alcoholics or debilitated patients are predisposed to *Klebsiella* or gram-negatives, and smokers or those with COPD are at risk for *Haemophilus.* Anaerobic infections may occur with aspiration or "postobstructive" pneumonias and can get necrotic and cavitate. Postobstructive pneumonia or collapse occurs distal to an endobronchial mass or narrowed, blocked bronchus. Antibiotic coverage should be adjusted according to the suspected bug type. X-ray improvement or decline may lag days behind the true clinical picture of pneumonia.

HIV + or immunocompromised patients get opportunistic infections like *Pneumocystis Carinii* pneumonia (PCP), which ~70% of AIDS patients eventually get (see Chapter 9) PCP is an endemic protozoan that is normally suppressed by the host's intact immune system. **PCP** can look like just about anything, but classically causes quickly developing, bilateral, interstitial, or dusty "ground-glass" infiltrate without effusions or adenopathy (see Fig. 9.2). The **ground-glass** pattern is an increased background haziness, as if someone crushed glass into very small particles.

Primary TB may have diffuse pneumonitis with hilar adenopathy and can occur anywhere. Postprimary or **reactivation TB** may involve the upper lobes and can cavitate (with a dark spot of air or necrosis within a nodule). If active TB is considered, TB masks should be worn until it is sorted out.

- "Friedländer's" lobar pneumonia has an expanded lobe, with outward bulging fissures (originally described with *Klebsiella*).
- "Atypical bacteria" (*Mycoplasma* and *Legionella*) can cause airspace and/or interstitial disease (as can tuberculosis).

OTHER LUNG DISEASES

In the setting of trauma, peripheral airspace disease that doesn't conform to lobar borders could be a **lung contusion,** which typi-

cally resolves in 3 to 5 days. With trauma, also look for rib fractures, pleural effusion (hemothorax), or pneumothorax (Fig.
2.15). Multifocal peripheral patchy airspace disease developing
in an ICU patient with fractures several days after trauma could
be fat emboli from the marrow.

Asthma, emphysema, **COPD,** or reactive airways cause air
trapping that makes the lungs look darker and increases the lung
volume, flattening the diaphragm. Peribronchial cuffing may be
seen in acute asthmatic attacks or bronchitis. Smoking or inhaled
particles can cause interstitial disease.

- "High-resolution CT" uses very fine 1-mm thin slices to subtype
  interstitial lung disease.
- Cystic fibrosis, silicosis, and TB cause upper lobe interstitial disease.
- Asbestosis, scleroderma, rheumatoid arthritis, and aspiration
  cause lower lobe interstitial disease.
- Alpha-1 antitrypsin deficiency causes lower lobe emphysema.

Solitary Lung Nodule

The **solitary pulmonary nodule** may be most commonly: a) a simple granuloma, b) an early resectable primary cancer, or less
commonly, c) a hamartoma, or d) an isolated met. The most
reliable way to tell neoplasm from benign lesion is growth or
change over time, but pattern of calcification may also help. **A
lesion that has not changed in size or appearance over 2 years
and is calcified is almost certainly a granuloma and requires no
further work-up,** biopsy, or follow up. Stippled, popcorn, particu-
late, circumscribed, or laminated calcification in a stable nodule
is benign. Calcium in the periphery of a nodule is indeterminant,
as this could also represent a "scar carcinoma" (adenocarcinoma) growing in an old granuloma. A new nodule needs a CT to
check for microcalcification, other nodules, and for adenopathy.
Peripheral nodules or masses may be biopsied through the skin
under CT or fluoroscopic guidance, which is less risky than open
surgery. Central masses may be approachable with bronchoscopy.
The remainder may need surgery or follow up. To best look for
nodules, **glance close to the film (inches), then step back (feet)
and look again.**

LUNG CANCER PEARLS

- **S**quamous cell and **S**mall cell carcinoma are **S**entral (central) lesions.
- Squamous cell carcinoma is highly-associated with smoking, can cavitate, is slow growing, has a male predominance, and spreads by lymphatics.
- **Pancoast's superior sulcus tumor** is squamous cell and may present with a Horner's syndrome (ptosis, miosis, anhidrosis).
- Small cell carcinoma grows fast, spreads by blood, is most associated with paraneoplastic phenomena, and has bulky hilar or mediastinal adenopathy.
- Alveolar cell carcinoma is a subtype of adenocarcinoma that can present as chronic airspace disease without a discrete mass.
- Adenocarcinoma presents peripherally or as airspace disease, and is also associated with smoking (possibly low tar/low nicotine cigarettes [inhaled deeper?]).
- "Golden's reverse 'S' sign" = A right hilar/suprahilar bulging mass with right upper lobe retraction and volume loss causing the minor fissure to bend like the letter "S" turned upside-down and on its left side.
- Lung mass doubling time < 1 or > 18 months is unlikely to be malignant. (That is double volume, not double diameter.)

PULMONARY EMBOLUS (PE)

PE may be the biggest preventable underdiagnosed killer of hospitalized patients. Always consider DVT prophylaxis. The most common chest x-ray findings include atelectasis, pleural fluid, or nothing at all. A less common finding is a lung infarct or "**Hampton's hump**" (Fig. 2.26). This is a triangular wedge density with the wide base at the periphery and a pointed angle towards the hilum or origin of blood supply. (Any organ infarct will tend to form a wedge with the point towards the hilum or blood supply [i.e., spleen, kidneys, brain].) An uncommon to rare finding with PE is "Westermark's sign," which is a focal decrease in perfusion (oligemia) causing relative lucency of one lung or lobe in relation to the rest of the lungs.

A PE is a clot that causes a "perfusion defect" with normal ventilation on a **nuclear medicine V/Q (ventilation/perfusion) scan** or a "filling defect" on **pulmonary angiography.** The initial

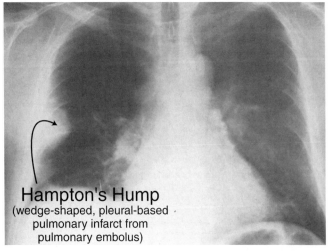

Figure 2.26. Hampton's hump.

study is the V/Q that involves a comparison of three pictures: chest x-ray, ventilation scan, and perfusion scan. Indications for angiography include a) indeterminant or intermediate probability V/Q scan, b) high suspicion for PE with a low probability V/Q, and c) high probability V/Q but high risk for anticoagulation.

If the patient has COPD, CHF, or infiltrates, then an indeterminant V/Q scan is more likely. Two or more lung segments that get air but not blood (ventilation-perfusion mismatch) without chest x-ray abnormalities (in that same area of the lung) make a scan high probability for pulmonary embolus (Fig. 2.27). Contrast CT may show PE as dark spots plugging the normally bright pulmonary arteries (Fig. 8.2). Helical CT may prove to be useful in screening for PE (controversial). Massive central PEs may qualify for thrombolytics (controversial). *The chest x-ray should ideally be taken within 12 hours of the V/Q scan (less if possible).*

SEGMENTAL ANATOMY

Segmental anatomy is important in ventilation/perfusion scan interpretation. The right lung has three lobes and the left lung

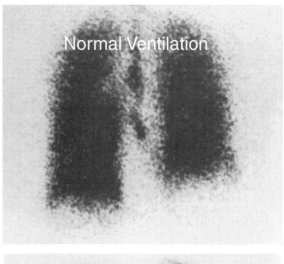

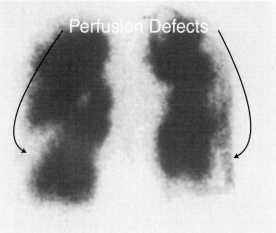

Figure 2.27. High probability V/Q scan.

has two. The lingula of the left upper lobe (LUL) lives in a roughly similar location to the right middle lobe (RML) on the lateral film. The lingula has superior and inferior segments, whereas the right middle lobe has medial and lateral segments. The right upper lobe (RUL) has apical, posterior and anterior

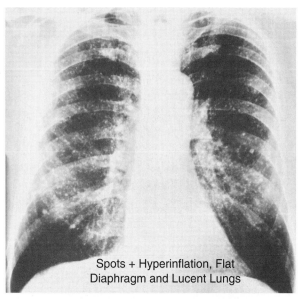

Spots + Hyperinflation, Flat
Diaphragm and Lucent Lungs

Figure 2.28. Miliary pattern and COPD

segments. The LUL is the same except for the fused apicoposterior segment and the presence of the lingula. The right lower lobe (RLL) has four basal segments (anterior, posterior, medial, and lateral) and one superior. The left lower lobe (LLL) is the same except for the fused anteromedial basal segment.

LATERAL FILM

Match and confirm findings from the PA with the lateral. The lateral film is often the best way to see hilar adenopathy, effusions, spine disease, and exact location of nodules or infiltrates. It also clarifies segmental anatomy and localizes findings.

Differential Diagnosis (DDX)

- DDX of single pulmonary nodule: granuloma, primary cancer, hamartoma, metastasis.
- DDX of "miliary" (diffuse small nodules): thyroid mets, TB, histoplasmosis, varicella, silicosis, sarcoidosis, pneumoconiosis (Fig. 2.28).

- DDX of cannonball mets: colon, renal cell, sarcoma, melanoma, testicular.
- Cavitary neoplasms: if primary, then squamous cell carcinoma > adenocarcinoma. If metastatic, then sarcoma met, or squamous cell carcinoma met (from head and neck in males or cervix in females).
- DDX of cavitary lesion : "**CAVITY**"—C = cancer, cystic fibrosis; A = autoimmune (rheumatoid arthritis, Wegener's); V = vascular (septic emboli); I = infection (bacterial or fungal pneumonia); T = traumatic pneumatocele, tuberculosis; Y = young = sequestration, bronchogenic cyst (Figs. 2.29 and 9.2).

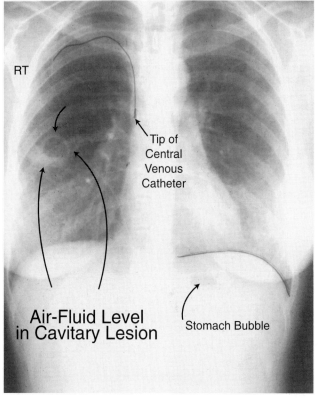

Figure 2.29. Cavitary lesion.

SUMMARY

When you think you're finished, look again at your personal "psychological blind spots." This should include at least the apices, CPAs, retrocardiac lung, retro-diaphragmatic lung, hila, and anywhere you've missed findings before. Never make the same mistake twice.

Remember to look for **ABCDEFGHI:** Airway, Bones, Cardiac, Densities, Effusions, Gastric bubble, Hila, and Inspiration.

Suggested Readings

Chen MYM, Pope TL, Ott DJ, eds. Basic radiology. New York: McGraw-Hill, 1996.

Dahnert W. Radiology review manual, 2nd ed. Baltimore: Williams & Wilkins, 1993.

Elliott LP, ed. Cardiac imaging in infants, children, and adults. Philadelphia: JB Lippincott, 1991.

Felson B. Chest roentgenology. Philadelphia: WB Saunders, 1973.

Freundlich IM, Bragg DG. A radiologic approach to diseases of the chest. Baltimore: Williams & Wilkins, 1992.

Ravin CE, Cooper C, Leder RA, eds. Review of radiology, 2nd ed. Philadelphia: WB Saunders, 1994.

Squire LF, Novelline RA. Fundamentals of radiology, 4th ed. Cambridge: Harvard University Press, 1988.

Weist P, Roth P. Fundamentals of emergency radiology. Philadelphia: WB Saunders, 1996.

Chapter 3

Neuroradiology

Part I   Basics

Nowhere in medicine have technologic advances in imaging had such an impact as in neuroradiology. The practice of neurology, neurosurgery, and emergency and trauma medicine has been radically transformed in the past 2 decades by magnetic resonance (MR) and computed tomography (CT) imaging of the skull, spine, and brain. Neuroradiology all boils down to one rule: abnormal things distort, disrupt, move, or enlarge normal central nervous system (CNS) anatomy (Figs. 3.1 to 3.5).

PLAIN FILMS

Plain skull and facial radiographs have a limited role in the trauma setting and are overused. These films are very busy looking, with many overlapping bone shadows (Fig. 3.6). **CT** is more sensitive, is easier to read, and is the gold standard. It provides fast and reliable evaluation of the skull and intracranial contents and has largely replaced skull and facial radiographs in the setting of trauma (what matters is what's in the egg, not the eggshell itself). CT should not be delayed to wait for the outcome of skull films, which rarely change management and may even delay surgery or treatment of trauma complications. Skull films may be helpful in suspected **child abuse,** along with a bone scan and a directed radiographic survey. Bone scan may show areas of hidden or old trauma. (The hallmark of child abuse is fractures of different ages—see Chapter 6.)

FRACTURES

"**Depressed**" (pushed-in) or displaced **skull fractures** that require surgery won't be missed on CT. Depressed fractures may

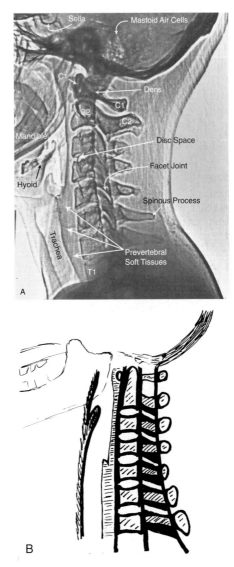

Figure 3.1. A. Lateral c-spine. B. C-spine alignment lines.

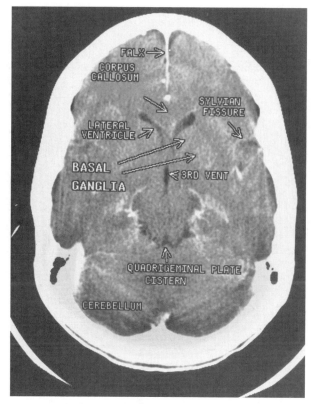

Figure 3.2.

overlie a rapidly expanding, extra-axial (outside the brain, inside the skull) hematoma, in which case surgical evacuation is a life saver. Temporal or skull base fractures can result in hearing loss, facial nerve injury, cerebrospinal fluid (CSF) leak, or carotid artery tear (dissection), and are best seen on CT. Likewise, CT depicts orbital "blowout" fractures best. Basically, skip the trauma skull film, because you will be treating your own curiosity more than the patient.

Orbital blowout fractures involve either the orbital floor (maxillary sinus roof) and/or the medial orbital wall (lateral wall

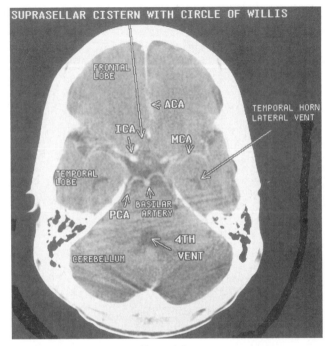

SUPRASELLAR CISTERN WITH CIRCLE OF WILLIS

FRONTAL LOBE

< ACA

ICA

MCA

TEMPORAL HORN LATERAL VENT

TEMPORAL LOBE

PCA BASILAR ARTERY

4TH VENT

CEREBELLUM

Figure 3.3.

of the ethmoid sinus, or "lamina papyracea"). The contents of the orbital cavity herniates into the adjacent sinus, sometimes entrapping muscle and leading to diplopia. The anatomy of the orbit is best depicted on coronal CT (Fig. 3.5). This can be directly acquired coronally or reconstructed by computer processing of axial images. The reconstructed images are useful in the trauma patient who can't extend the neck for coronal CT because of an unstable or incompletely evaluated c-spine. Gas in the orbit and blood in the maxillary or ethmoid sinuses are hallmarks of orbital blowout fractures (Fig. 3.7). Blowouts might show the horizontal line of a blood-air level in a maxillary sinus or a dense sinus completely full of blood (opacified sinuses are also seen with sinusitis).

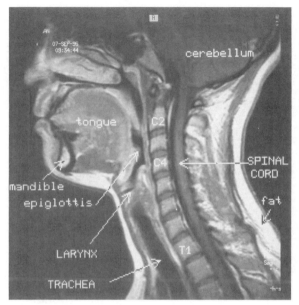

Figure 3.4. MRI.

PHYSICS

The bottom line is that CT shows clinically important fractures. Rarely, however, a skull film will show a horizontal fracture that may not be seen on CT because it is in the imaging plane. This is a fine example of "**volume averaging,**" where the dark fracture line and the white bone are contained within the same slice and are averaged together, resulting in white density (dark + very-white = less bright, but still white). This is rarely a problem in this setting, but can limit evaluation of structures that are smaller than the CT slice thickness. Also, beware of normal vascular grooves and suture lines that may simulate nondisplaced fractures.

The posterior fossa is also prone to "streak" and "beam hardening" artifacts related to dense bone in the skull base. Metals in dental work or surgical clips also cause a "streak" artifact that

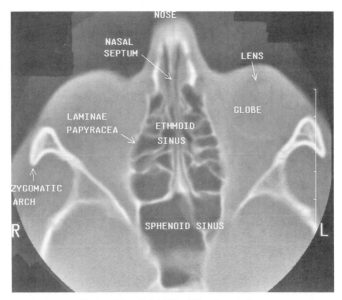

Figure 3.5. Axial CT: orbits, sinuses, face.

appears as radiating bright lines in a starburst pattern. The x-ray beam may also become "hardened" and more penetrating as it goes through more tissue due to selective filtering of the weaker x-rays. OK, enough physics . . .

C-SPINE

Cervical spine (c-spine) evaluation must be firmly founded on clinical examination and history of mechanism. After blunt multisystem trauma, the c-spine must be immobilized until injury can be excluded by a combination of clinical and imaging evaluation. No c-spine film should be ordered if the minor trauma patient has no neck pain, no neurologic findings or mental status changes, no mechanism by clear history and no other complicating or distracting injuries. C-spine exams for documentation are overused.

Treat the patient and not yourself or your fear of litigation. Be inquisitive and do not copy unproven practice habits that do not make sense; however, it is often better to err on the side of

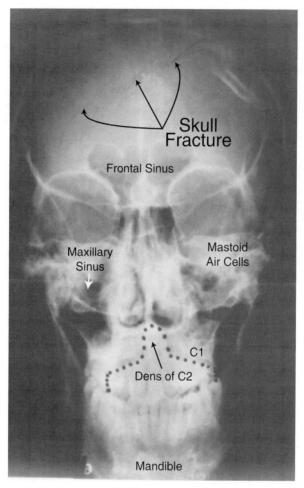

Figure 3.6.

caution. If you take the time to debate whether or not to order films, be safe and get the information. Deciding what studies to skip is the challenge that defines good clinical judgement. OK, enough philosophy . . .

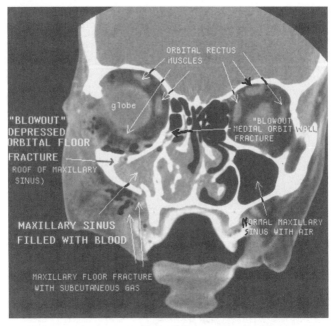

Figure 3.7.

C-SPINE COOKBOOK

Technique:	Includes C7–T1?
Alignment:	Listhesis? Jumped facet?
Fractures:	Stable or unstable?
Soft tissues:	Prevertebral swelling?
C1–C2 Relationship:	Pre-dens distance?
More studies needed:	Flexion/extension, CT, MR

The c-spine series begins and ends with the lateral, as most abnormalities will be seen or suggested on this one view (Figs. 3.1 and 3.8). **Learn normal,** spend time with it, become friends with it. In the trauma patient with an immobilized neck, the "**cross-table lateral**" screening exam is shot with the patient still in the collar, which cannot be removed until the c-spine is "cleared" radiographically *and* clinically. Check for the hospital's

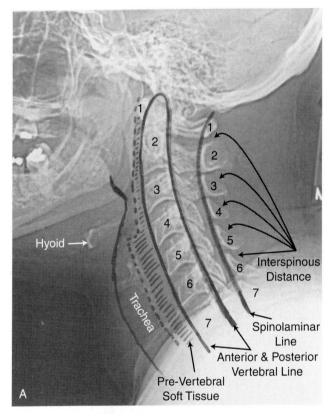

Figure 3.8. A. C-spine lines.

protocol. Clearing a c-spine requires seeing all 7 cervical vertebral bodies including the bottom of C7 and, ideally, the top of T1 (controversial). This is harder than it sounds, because of short necks and broad shoulders (Mike Tyson's neck would be a nightmare to x-ray). An attempt to see C7–T1 may be made with a "swimmer's view," where the arms are lifted up out of the way of the C7–T1 level. Alternatively, digital imaging with postprocessing can help see C7–T1 with a single exposure by rewindowing the image. CT may be necessary to clear C7 to T1.

Like the chest x-ray, the lateral c-spine lends itself to a cookbook checklist. An organized methodical approach is a must. Save

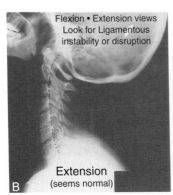

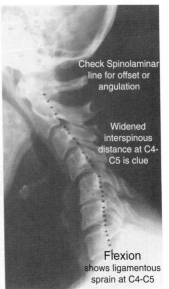

Flexion • Extension views
Look for Ligamentous
instability or disruption

Extension
(seems normal)

B

Check Spinolaminar
line for offset or
angulation

Widened
interspinous
distance at C4-
C5 is clue

Flexion
shows ligamentous
sprain at C4-C5

C

Trachea   Spinolaminar
       Prevertebral   Line
       Soft Tissue

Figure 3.8.   *(continued).* B. Flexion extension views. C. C-spine lines.

the bones for last. Look for foreign objects and tubes and lines first. Next, look for **soft tissue swelling,** especially in the "**prevertebral**" (in front of the vertebrae) or retropharyngeal space. This distance is mentally measured from the back of the airway to the front of the spine. One easy way to remember the maximum distance here for an extended neck is "3 × 7 = 21"; or 7 mm at C3, and 21 mm at C7. Another maximum normal measurement is one-half of the anterior to posterior (AP) vertebrae width for the upper half of the cervical spine. Now forget these numbers . . . they are very rough estimates, and should not be relied upon. Swelling can be seen near a fracture, bleed, mass, infection, or ligament injury; CT, MR, or history may help to answer exactly which. Get a "feel" for what a normal lateral looks like. Spending 5 minutes flipping through a basic text for pictures will be time well spent.

At this point, bite into the meat of the film and mentally draw the three straight lines naturally formed by the vertebrae and spinal canal (**anterior and posterior vertebral and spinolaminar lines**) (Figs. 3.1B and 3.8). These should be nearly smooth without sudden jumps or bumps. A jagged line or discontinuity may mean fracture or ligamentous instability and may require CT (these lines are easier for the eye to draw if you tilt the film almost flat and look obliquely up the patient's spine): A smooth bump from a bony overgrowth continuous with the spine (osteophyte) is likely a sign of **degenerative disc disease** (like osteoarthritis of the spine), especially if the underlying vertebrae remain well aligned, and there is also sclerosis and disc space narrowing.

Next, take a mental measurement of the horizontal distance between C1 and C2 on the lateral film, from the front arch of the C1 ring to the finger-like dens (odontoid) of C2 that normally sits just behind this. Normal is less than 3 mm in adults (or less than 5 mm in children who naturally have more elastic transverse and alar ligaments). An increased gap is diagnostic of **atlantoaxial** (C1–C2) **instability** (Fig. 3.9). Patients with rheumatoid arthritis or Down syndrome can have such loose ligaments here that the medulla or cord can be injured when the neck is forcibly extended (as for tracheal intubation).

If there is suspicion of ligamentous instability (but no fracture or neurologic deficit), **flexion and extension lateral** views can be compared to the neutral lateral to see if the vertebrae slide for-

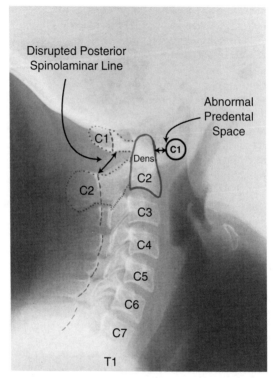

Disrupted Posterior
Spinolaminar Line

Abnormal
Predental
Space

C1

Dens

C1

C2

C2

C3

C4

C5

C6

C7

T1

Figure 3.9.

ward (antero**listhesis**) or backward (retrolisthesis) with voluntary motion. A trauma patient's neck should *never* be forcibly moved, even cautiously. If any questions remain or if the patient is unconscious, CT or even MR is the next step.

C-spine CT is indicated for severe head injury, severe pain, neurologic deficits, inadequate or equivocal plain films, high suspicion of neck injury, or for more exact anatomic depiction of fractures. CT is playing an increasing role in trauma c-spine evaluation, especially with helical CT, which can image the whole c-spine in seconds. The more detailed sagittal reconstructions possible with helical CT appear similar to a lateral plain film (Fig. 3.13).

The complete **c-spine series** includes an AP, lateral, AP "open-mouth dens," and, possibly, oblique films. The AP looks for rotational instabilities (and fulfills the "two views at 90 degrees" rule for bone fractures). The controversial oblique films better display the facet joints, neural foramen, and the posterior elements (laminae, pedicles, and pillars). The AP open-mouth dens view shows the C1–C2 relationship, and the lateral masses of C1 and C2 should line up on this view. If C2 extends wider than C1, consider a "**hangman's fracture**" of C2. (If they are offset greater than 7 mm, the ligaments are likely disrupted as well.) A hangman's fracture involves both neural arches of C2 (Fig. 3.10). ("Harris' ring" should normally be a dense continuous ring overlying C2 at the base of the dens on the lateral film. Ring disruption may be the only subtle clue of a nondisplaced hangman's fracture.)

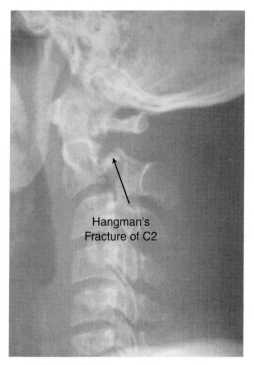

Hangman's
Fracture of C2

Figure 3.10.   Hangman's fracture of C2.

Also look for the horizontal base of the **dens fracture** (type II dens fracture) between the dens and the body of C2. This can be mistaken for an "os odontoidium," an ossification variant (which can also have ligamentous instability and require fixation). Another fracture simulator is the not-yet-united dens ossification center in children. These predominantly horizontal fractures and variants can be missed on transverse CT (due to "volume averaging"), but are usually caught if sagittal reconstructions are performed.

SPINE AND DISC DISEASE

Spondylolysis is a usually asymptomatic defect in the pars interarticularis that may be a fracture or may be congenital. A neck collar on the "scottie-dog" on oblique spine films is diagnostic (usually L4 or L5) (Fig. 3.11). It allows the vertebral bodies to slide forward or back on each other (**listhesis**). This has the unfortunate name of spondylolysis spondylolisthesis.

Disc herniation (or herniated nucleus pulposis) may push on the spinal cord, but more commonly presses on or impinges adjacent nerve roots (Fig. 3.12). A common question is: which nerve root is squeezed by an L4–L5 disc herniation (or bulge, protrusion, or extrusion)? The typical radiologist answers . . . "it

Figure 3.11. Oblique lumbar spine film: Scottie dog collar = spondylolysis.

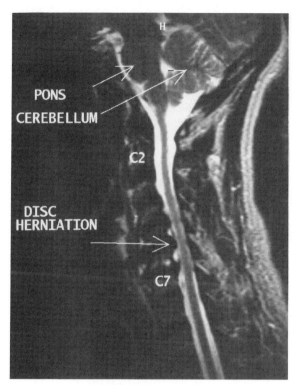

Figure 3.12.   MRI C-spine.

depends." (People want to hear black and white answers, but board questions and patients are made with a lot of gray.) A **lateral** L4–L5 disc herniation may cause reflex and sensorimotor deficits in the L4 distribution, while a **central** (dorsal) or nearly central (paracentral) one may cause L5 findings. Think anatomically to remember this: the higher root (L4) leaves the canal out the lateral neural foramen, while the more inferior root (L5) continues downward, near midline (so it can dive out lateral just below this).

Occasionally, **CT myelography** may also be necessary to clarify complex postoperative disc cases. This seemingly barbaric study requires a lumbar puncture with injection of contrast into the spinal sac (intrathecal) followed by CT scan. Today, MR has nearly replaced CT myelography, and is the study of choice for initial disc, spinal cord, ligamentous, soft tissue, epidural and nerve root evaluation.

A common indication for emergent **MR** is suspicion of acute **cord compression** from trauma, metastases to the spine or para-vertebral tissues, or epidural abscess or hematoma. This presents with neurologic deficits and/or bowel or urinary incontinence. Before sending the patient with possible cord compression to the lonely MR with limited monitoring and hours of delay, **give steroids** to decrease swelling. This is also true for CNS swelling from brain metastases with pending brain herniation. Cord compression may need urgent surgical decompression or emergent radiation therapy.

In the trauma patient with any type of c-spine study, remember to check for the basic "**ABCs**" of c-spines: **A**lignment, **B**one, **C**ord, and **S**oft tissue swelling.

PLAIN FILM TRAUMA PEARLS

• Jefferson fracture = C1 ring breaks in at least two places (a pretzel doesn't break in only one place; this principle also applies to the bony rings of the pelvis). Mechanism: an axial load, like diving into an empty pool.
• Clay shoveler's fracture = spinous process avulsion (usually lower C-spine).
• Teardrop fracture = avulsion with a chip off the anterior corner of a vertebral body attached to the anterior longitudinal ligament. Mechanism: flexion or extension.
• Jumped facet = locked facet = roof shingle of the facet disarticulates (dislocates) and jumps over the facet below and commonly also fractures. Stable if unilateral, unstable if bilateral, but *treat both as unstable*. A half-jumped facet sitting on top of the facet below is a "perched facet." On transverse CT, an appearance like a sliced hamburger bun is a normal facet, and an appearance like upside-down hamburger bun halves is a jumped facet.
• Burst fracture = vertebral body explodes from axial load. One can pop a bone fragment posteriorly into canal (retropulsed fragment) (Fig. 3.13).

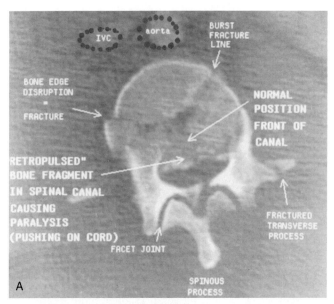

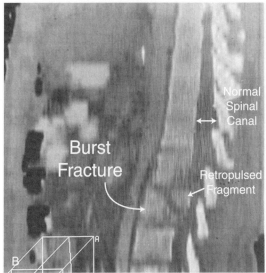

Figure 3.13. A. Axial CT of burst fracture. B. Sagittal CT reconstruction.

- Compression fracture = vertebral body squashed. More commonly wedged with loss of height in the front. Look for paraspinal hematoma (if acute), dense bony sclerosis (if chronic), or old films to assess age. Etiology: trauma, multiple myeloma, metastases, or osteoporosis (Fig. 3.14).
- Chance fracture = usually upper L-spine horizontal fracture. Involves vertebral body and posterior elements. Mechanism: hyperflexion waist. Also called "lap seat belt fracture."
- Unstable c-spine fractures include: Jefferson's, hangman's, teardrop, bilateral jumped (locked) facets.

Part II Trauma and Blood

CT has revolutionized emergency radiology, especially of the brain and abdomen. MR provides exquisite detail but is more expensive than other modalities, and should be used only if it could change your management or help your patient in some way. Make sure there is a question you want answered before you order a test. As imaging advances, often more information is obtained than we know what to do with. Order an MR (or any test or lab study) only when you clearly understand the indications and usefulness. Incidental findings often lead to unnecessary studies, procedures, complications, and worry.

CT and MR provide a vital window to CNS pathology. The CT findings in a stroke or trauma patient may be predicted based on the physical examination or clinical presentation. Focal neurologic findings often point to the lesion location. Now to more physics-made-stupid . . .

PHYSICS

In plain x-rays and CT, brightness is a measure of tissue density (or how many x-rays it stops, or "attenuates"). **Metal is more dense and thus brighter than bone > blood > brain > cerebrospinal fluid (CSF) > fat > air.** MR images water or proton content (the density of water in tissue). MR uses this fact to locate and quantify abnormal tissue, edema, swelling, and blood–brain barrier breakdown. MR brightness depends on the type of MR radiofrequency pulse sequence. Water and CSF are bright on T2 imaging, whereas fat is bright on T1 sequences. (Remember H2O is

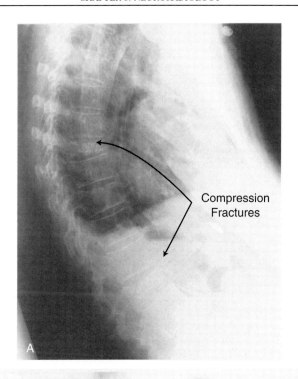

Compression
Fractures

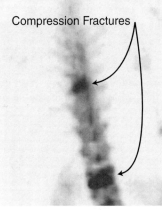

Compression Fractures

Figure 3.14. A. Lateral thoracic spine. B. Bone scan.

bright on T**2** images). T1 images better define the normal anatomy, and T2 images better define pathology.

HEAD CT COOKBOOK

Blood:	SAH, EDH, SDH?
Bones:	Fractures?
Mass effect:	Swelling or midline shift?
Extra-axial:	Fluid collections?
Ventricles and sulci:	Size & balance? Hydro?
More studies needed:	IV contrast or MR?

CT OF HEAD TRAUMA

Immediate **unenhanced head CT** is needed in the acute trauma
patient with neurologic deficit, change in mental status, loss or
decreased level of consciousness, or difficult evaluation (alcohol,
drugs, poor history, dementia). Most after-hours or emergent
head CTs are for trauma or mental status changes, and almost
all can be done without contrast; however, contrast CT occasionally adds some information. When the diagnosis or history is un-
clear, do both unenhanced and contrast enhanced studies. Contrast may light up an unrecognized lesion, tumor, or infarct.

CT has major limitations in the **posterior fossa** (cerebellum,
midbrain, pons, and brainstem). If you suspect posterior fossa
pathology (vertigo, ataxic gait, cranial nerve findings, loss of consciousness, hypertension), order CT with "thin posterior fossa
cuts." These are thinner slices to try to decrease the artifacts due
to the thick bones of the skull base. If readily available, MR is
better here.

If there is facial trauma, or if the c-spine could not be cleared
on plain x-ray, these areas can be CT scanned at the same time.
A limited cervical CT directed at the level in question may be
ordered. Of course, the neck should remain immobilized if there
is any suspicion of neck injury.

CT TRAUMA PEARLS

• On a trauma head CT, look at the soft tissue windows first; a
  superficial hematoma may be a clue to location of underlying
  injury.
• On a trauma facial CT, if there is no blood in the sinuses and

no zygoma or nasal fractures, then there are no facial fractures (communication from Stuart Mirvis, study in progress).

- Simplified **Le Fort** facial fracture types: I = maxilla and alveolar ridge (with mobile palate); II = I + pterygoid plates and medial orbit (with mobile maxilla); III = II + zygomatic arch (with mobile face).
- The # 1 facial fracture is the zygomaticomaxillary complex fracture = "**tripod fracture**" = zygomatic arch, orbital process, and maxillary process fractures.
- $\frac{1}{3}$ of orbital blowout fractures involve *both* the floor and medial wall of the orbit.
- "Cephalohematoma" = subperiosteal blood that occurs outside skull (outer table), rather than inside skull (inner table), as with epidural hematoma.
- Intracranial air may signify sinus or mastoid fracture.
- Isodense subacute subdural hematomas are hard to see (1 to 3 weeks).

BLEEDS

Bleeds come in five flavors: **epidural, subdural, subarachnoid, parenchymal, and intraventricular.** An *epi*dural hematoma (outside the dura) is less common than a subdural hematoma (inside the dura), but both are extra-axial (outside the brain, inside the skull). Epidurals are usually from temporal-parietal fractures with tearing of the middle meningeal artery. This high-pressure leak rapidly expands in the epidural or subperiosteal space and forms a lens-shaped collection of brightness (high attenuation) on a CT without intravenous contrast (Fig. 3.15). This biconvex epidural lens classically does *not* cross suture lines because it is tagged down by strongly adherent periosteum. However, it *can* cross midline.

This is contrasted to the thinner, crescent-shaped *sub*durals that are usually caused by slow oozing of torn bridging cortical veins (Fig. 3.16). Subdurals *can* cross suture lines but do *not* cross midline; however, if bilateral, they may appear to cross midline. Subdurals are easily missed without blood windows.

The classic history for an epidural is loss of consciousness (LOC) followed by a lucid interval then decreased consciousness again. Subdurals may classically occur in alcoholic or elderly de-

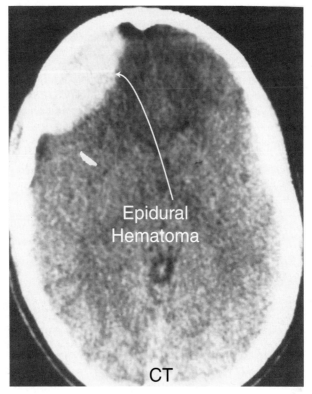

Epidural
Hematoma

CT

Figure 3.15.

mented patients following a fall. Atrophic, smaller brains don't quite fit in the larger skull, and have more potential to roll around in the sudden deceleration in falls. Also, tentorial or falx subdurals can be seen in infant abuse (**shaken baby syndrome**) (see Fig. 6.1). Infant abuse can also show shear injuries, coup-contracoup contusions, and retinal bleeds. A subdural "hygroma" is a subdural collection of CSF and may look just like a chronic subdural hematoma. It usually occurs in very young or very old patients.

On CT, subdural hematomas are classically bright during week 1, the same density as brain (isodense) during weeks 2 and 3, and become dark (hypodense) at around 3 weeks. There is

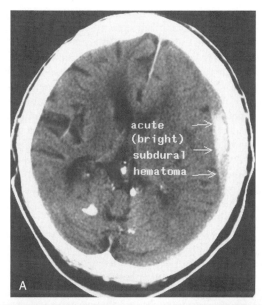

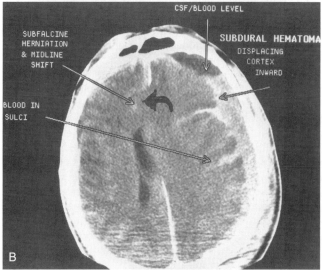

Figure 3.16.

considerable variation here, however, and they can rebleed. A CT or an MR with contrast may help to see the enhancing membranes of an isodense subacute or chronic bleed. Subdural collections can also develop if too much fluid is taken out of the ventricles too rapidly following ventriculoperitoneal shunting. Similar phenomena may cause pulmonary edema to develop following rapid removal of large amounts of pleural fluid, or cause low blood pressure following massive peritoneal fluid aspiration.

Subarachnoid hemorrhage (SAH) can be from trauma, ruptured aneurysm, arteriovenous malformation, adjacent brain bleed, or systemic coagulation problems. Of the three types of outside-of-the-brain bleeds, this blood is closest to the brain. It may be very subtle and can be easily missed in the on-call setting. The classic history for ruptured aneurysm SAH is in middle-age, with sudden onset of "the worst headache ever." It is also the most common type of bleed with head trauma. On unenhanced CT, SAH has brightness in the cisterns, sulci, and interhemispheric fissures where CSF lives. This manifests as bright finger-like projections interdigitating between gyri (Fig. 3.17). A SAH next to the falx at midline may look just like a bright calcified falx, except the edges of blood are blurry and ill-defined, whereas calcium has a smooth, well-defined border. Blood and bone windows may also help distinguish blood and calcium (compare to bony skull). If there are no other signs of blood, a bright falx or tentorium is likely normal.

Intraventricular bleeds occur in approximately 5% of head trauma patients and show up as brightness within the ventricular system, often an extension of an adjacent parenchymal bleed. Bleeding in the ventricles also can occur in premature neonates, although the germinal matrix (which lives in the caudate-thalamic notch) is a more common location in these infants.

Coup-contracoup contusions are traumatic **intraparenchymal** (inside the meat of the brain) bleeds, resulting from blood oozing from the direct hit as well as the "snap back" force on the exact opposite side of the brain. This classically occurs in the subfrontal or subtemporal region (coup) and in the opposite occipital area (contra-coup). **Contusions** are the most common posttraumatic parenchymal bleeds, whereas hypertension is the most common cause of brain hematoma (see part III—Nontraumatic, Stroke section). The CT appearance can be dark, bright or mixed, and usually has surrounding edema (Fig. 3.18).

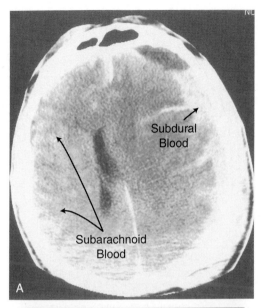

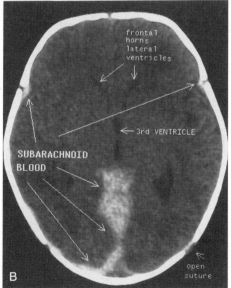

Figure 3.17.

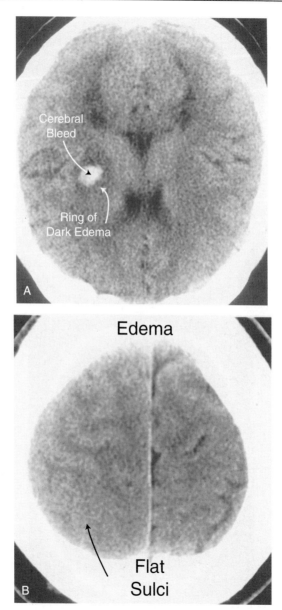

Figure 3.18.   A, B. Bleed with edema (flat sulci).

Posttrauma shearing or "diffuse axonal" injury results in tearing at the grey-white junction and multiple, small bleeds in the white matter below the cortex. These can be hard to see with CT and are better seen with MR. This diffuse axonal shearing injury is usually associated with immediate LOC.

EDEMA

Focal **edema** (or swelling with increased water content) may be seen near any CNS lesion or bleed as local darkness on CT, or brightness on MR T2. Asymmetry is often the only clue to diffuse hemispheric edema. This is often best seen on the highest axial CT slice near the vertex which shows a lack of sulci and gyri (flattened or "effaced" sulci) on the affected side (Fig. 3.18). This highest slice contains only the grey matter edge of the brain (the peel of the orange), thus better showing edema (lack of wrinkles in the orange peel).

Edema, swelling, or "**mass effect**" may cause **midline shift,** where the falx, corpus callosum, and third ventricle move away from the swollen hemisphere (Fig. 3.17). Even a small midline shift may be a bad prognosticator and may precede subfalcine (under the falx) herniation. Diffuse bilateral edema first flattens the sulci and can later make the ventricles small. However, this may be difficult to see without serial CT's showing a change from baseline.

CT VERSUS MR

MR is more expensive and slower than CT, and has been less sensitive for acute blood and SAH in the past (controversial). Patients with certain ferromagnetic surgical devices, some old brain aneurysm clips, all pacemakers, or metal fragments should *not* be placed in the magnet, because these objects could move in the MR (check operation notes and call MR). MR is also very sensitive to motion artifact (because longer "exposure times" than CT may yield blurred images). MR is usually less accessible to the emergency team and equipment than is CT. CT evaluates bone and calcium better than MR, and is much more sensitive for fracture (although MR sees fracture complications as well as, or better than CT). MR is superior to CT for reconstructing in different planes, for detection of nonbloody brain injury such as edema, for shear injury, for soft tissue injury, for posterior fossa evaluation, and for small lesions and nonacute bleeds. MR is also more sensitive to small concentrations of contrast than CT. Overall, however, CT continues to play a more important role in the emergency and on-call setting. Many of the principles of anatomy and enhancement are similar for CT and MR. If you have to pick one to learn well, make it CT.

MR BASICS

- MR "**T1**" images: fat is bright. T1 is best for CNS **anatomy.** Use T1 with gadolinium IV contrast to check for enhancement and blood–brain barrier breakdown.
- MR "**T2**" images: water (edema, interstitial or cellular fluid, cysts) is bright (**T2** for H_2O). T2 is best for most CNS **pathology.**
- MR "proton density" images may differentiate CSF from edema or water. Good for multiple sclerosis plaques or transependymal resorption of CSF in hydrocephalus.
- Flowing blood is dark on MR T1 and T2 (spin echo) images, and bright on "gradient echo" images.
- Nonmoving extravascular blood (hemorrhage or hematoma) is bright or dark depending on the blood age and type of MR image "sequence" (T1, T2, proton density, gradient echo). Usually, different age blood exists simultaneously. Think of blood of different ages any time you see a swirled or funky collection of bright and dark (also consider a rebleed).
- Very old blood (hemosiderin) is dark on nearly all sequences.

  Bleeds age from the outside-in; hemosiderin forms dark peripheral rings at the edge of bleeds.
- Learn basic principles of T1 and T2 sequences (what is bright or dark?). Ignore the rest of the sequences for now.

Part III   Nontraumatic

ANEURYSMS AND SAH

Nontraumatic sources of brain bleeds commonly include aneurysms and uncontrolled hypertension. Less common causes include arteriovenous malformation (AVM), bleeding coagulopathy, tumor, or hemorrhagic infarct. **Aneurysm** rupture results in **SAH,** which can sometimes help localize the aneurysm leak. For

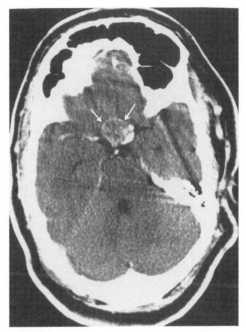

Figure 3.19. Berry aneurysm in suprasellar cistern.

instance, spontaneous SAH located mostly in the anterior inter-hemispheric fissure may suggest that an anterior communicating artery (A COMM) aneurysm has ruptured. Congenital "berry" aneurysms are the most common type in the head and commonly occur where Circle of Willis vessels branch (Fig. 3.19).

The three most common aneurysm places are the A COMM, posterior communicating artery (P COMM), and middle cerebral artery (MCA). SAH may be complicated by rebleeding (approximately 40% by 6 months), **hydrocephalus** (big ventricles from blood products blocking CSF resorption in the arachnoid villi), or vasospasm (narrowing of vessels which can decrease blood supply enough to cause infarcts approximately 4 to 14 days post-SAH). Vasospasm may be treated with calcium channel blockers or balloon angioplasty, where a balloon is blown up on a catheter in a narrowed vessel. Like AVMs, some aneurysms are embolized with clot-forming metal coils. Embolization and angioplasty are performed by interventional radiologists (see chapter 8).

Lumbar puncture (xanthochromia and RBCs) and MR are more sensitive for SAH than is CT. Aneurysms can be seen with conventional angiography (requires invasive catheterization), MR angiography (MRA; not as reliable and less information), and even with CT, if big enough.

Types of Aneurysms

Aneurysms are either "fusiform" (tubular dilation) or "saccular" (discrete outpouching). **Pseudoaneurysms** bulge out but don't contain all three layers of vessel wall. "False" aneurysms are actually ruptures contained by surrounding soft tissues and don't have vessel wall at all (see Chapter 8). In the head, we're dealing with congenital saccular aneurysms usually. Other kinds of less common aneurysms include mycotic (from infectious emboli, usually in the peripheral cortex or grey-white junction), traumatic (cortex near falx), atherosclerotic, or hypertensive ("Charcot-Bouchard"; small aneurysms in the perforating vessels that can lead to hypertensive bleeds).

Hypertensive Bleeds

The most common cause for an intracerebral hematoma is hypertension. These bleeds are most commonly found in the putamen of the **basal ganglia** (#1 place), **pons,** and dentate nucleus of the **cerebellum,** in order of likelihood (common boards question).

STROKE

Suspected **stroke** or cerebrovascular accident (**CVA**) is one of the most common indications for emergent head CT. In an **ischemic**

CVA, brain cells may die (infarct) from not getting enough blood and oxygen. This may lead to swelling, edema, and mass effect from excess water and exudate between and in cells. An **acute CVA** (less than 24 to 48 hours) may show edema in a vascular distribution that shows up as a vague, darker decreased density on CT, with loss of sulci definition and flattening of the brain folds (Fig. 3.20). A less common finding in acute CVA is the bright line of an acutely clotted vessel. Usually, the CT is normal with less than 24 hours of ischemia.

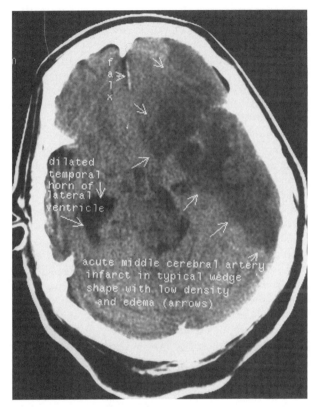

Figure 3.20.   MCA infarct.

Subacutely, the darker region of infarction later manifests. The mass effect and swelling maximizes at 3 to 6 days when the danger for herniation may be greatest. Infarcts may enhance with contrast at approximately 2 to 4 weeks. Chronic changes from old infarcts include scarring (gliosis) and volume loss or brain melt (encephalomalacia), which is dark on unenhanced CT (from CSF replacing the melted brain).

Causes of CVA are numerous and diverse. **Atherosclerosis** can result in narrowing (stenosis) or clots (stationary thrombi or mobile emboli from the heart or carotid arteries). Emboli tend to stop in the MCA territory. A truly **hemorrhagic infarct** suggests an embolic etiology; however, most **embolic infarcts** do not bleed (pale infarcts). Anticoagulants are contraindicated if the infarct is bloody or very edematous. Alternatively, thrombolytics may help if an ischemic pending infarct is superacute (symptoms less than 4 to 6 hours), and is not bloody or edematous.

Atherosclerosis and hypertension can impair the blood flow in small penetrating arteries to the deep matter, resulting in small infarcts in the basal ganglia, thalamus, internal capsule, brainstem, and periventricular white tracts. These "**lacunar**" **infarcts** are seen chronically as little dark holes. If this results in dementia, it is called "Binswanger's" dementia (not to be confused with **multi-infarct dementia,** which is usually from cortical infarcts).

In the ER setting, **hypoxic infarct** can be seen from blood clot, heart failure, shock, blood loss, drowning, or carbon monoxide poisoning and tends to involve areas where two different cortical arterial distributions meet, known as "**watershed**" **infarct.**

A young patient with a neurologic deficit and recent weight lifting, strenuous activity, or motor vehicle collision may have a carotid or vertebral "**dissection.**" (Dissection is blood entering the media of a artery wall through a torn inner intima and possibly running or clotting in this "false lumen," parallel to the true lumen.) This can be diagnosed with MRA or conventional angiography.

CVA can also result from **vasculitis** associated with collagen vascular diseases (lupus, polyarteritis nodosa, temporal arteritis), radiation therapy, meningitis, or drugs (especially cocaine and speed). MRA and angiography show the irregular inflamed vessels and CT and MR show the end product of these insults. Venous clot and vasospasm after SAH are less common causes of CVA.

MR is more sensitive than CT for acute ischemia; however, it has limitations in the emergency setting. Conventional angiography is more useful than MRA for defining quantity and quality of vessel clots, ulcerations, or dissections, as well as how many aneurysms are present and which ones bled. MRA may overestimate vessel narrowing. CT usually answers the immediate questions "Why the CVA? How old is it? What can be done now?" and is therefore the most useful exam in the emergency and on-call settings.

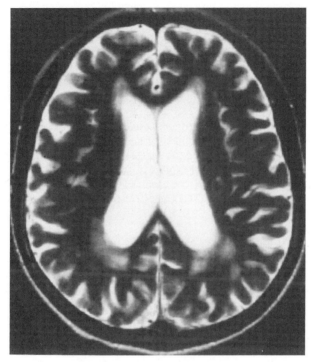

Figure 3.21.   T2. MRI hydrocephalus.

HYDROCEPHALUS

Big ventricles are seen in **hydrocephalus** (Fig. 3.21). As the normal brain ages, atrophy makes the ventricles, sulci, and cisterns slightly bigger. Thus, when considering atrophy, hydrocephalus, or edema, your sensitivity must be adjusted for the patient's age. Hydrocephalus is divided into obstructive and nonobstructive (communicating) causes. **Nonobstructive hydrocephalus** has no blockage and patients present as "wet, wobbly, and weird," (or incontinent, ataxic, and with dementia).

CSF PLUMBING

CSF flows from the choroid plexus in lateral ventricles through the foramen of Monroe to the 3rd ventricle through the aqueduct (of Sylvius), to the 4th ventricle through the foramen of **M**agendie (**m**idline) and **L**ushka (**l**ateral) to the cisterns and sulci, to finally be resorbed in the arachnoid granulations. This plumbing pathway may help localize the blockage, because different levels of blockage will result in characteristic patterns of hydrocephalus. For example, if the pathway between the 3rd and 4th ventricles is blocked, the lateral and 3rd ventricles get bigger, but the 4th ventricle remains small.

This hydrocephalus from blockage (**obstructive hydrocephalus**) may happen with congenital aqueduct stenosis, brainstem glioma, or post-SAH blood products clogging the arachnoid villi. Ventriculoperitoneal CSF shunts treat hydrocephalus, but may overdrain or underdrain. Overdrainage results in the "slit ventricle" syndrome with small ventricles, chronic subdural fluid collections, and headaches.

CISTERNS AND VENTRICLES BASIC PEARLS

• The lateral ventricle anatomy is simulated by the index finger and thumb (with the thumb pointing forward). The index fingers = temporal horns, the thumbs = frontal horns, and the wrist = the atria or posterior horns.
• The "suprasellar" cistern is above the sella and rhymes with "stellar" and conveniently is also star or "stellar" shaped, with five points. (Just call it the "suprasTellar" cistern.) The posterior cerebral arteries (PCAs) live in the rear points of the star; the MCAs live in the side points; the anterior point houses the parallel anterior cerebral arteries (ACAs), with the short A COMM between them.
• The "quadrigeminal plate" cistern is shaped like the mouth of a "smiley face" and should always be smiling (Fig. 3.2). If it gets small and you see no smile, it may be because of uncal or transtentorial herniation.
• The ambient (or perimesencephalic) cistern is just lateral to the smile.
• The lateral and third ventricles form the eyes and nose of the "smiley face," (Fig. 3.2) communicating with each other via the "foramen of Monroe."

CLINICAL PEARLS

• Head CT is often indicated before lumbar puncture to exclude swelling or mass effect, which increase the risk of tap-induced herniation.

- Contrast enhancement usually signifies hypervascularity or breakdown of the blood–brain barrier.
- The most common intracranial neoplasms include metastases, gliomas, and meningiomas (Fig. 3.22). **Metastases are more common than primary brain tumors in adults.**
- **CNS abscess** commonly results from direct extension from sinusitis or mastoiditis, which show opacified air cells or sinuses.
- **Sinusitis** fills in the normally lucent air-filled sinuses. Acute sinusitis may have air-fluid levels (seen on waters view or on CT). Waters view is AP with the nose slightly turned up (so as not to breathe water while swimming?).
- **Discitis** = infected disc. Loss of endplate definition in two contiguous vertebral bodies is the earliest plain film finding. Eventually, paraspinal mass (abscess) and loss of vertebral body height occur. MR and bone scan are positive earlier. #1 bug is staphylococcus.

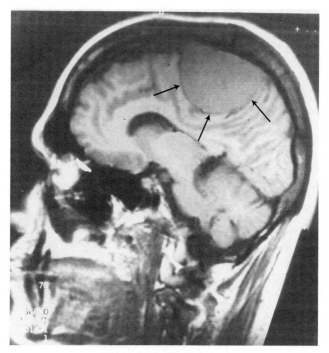

Figure 3.22. Meningioma.

- Meningitis may have meningeal enhancement.
- Herpes encephalitis involves the temporal (and frontal) lobes.
- Coarctation, polycystic kidney disease, and nearby cerebral arteriovenous malformations have an increased incidence of cerebral aneurysms.
- If an enhancing brain mass crosses midline (invading the corpus callosum), then it is a glioblastoma multiforme (an aggressive primary brain cancer) or lymphoma.
- Darkness on CT *next* to the lateral ventricles in the white matter may be due to age-related or ischemic-related changes from small vessel disease ("periventricular leukomalacia"—a common finding). Alternatively, this may represent "transependymal resorption of CSF" in hydrocephalus.
- The cavernous (venous) sinus contains cranial nerves III, IV, V_1, and VI, as well as the carotid artery. Carotid to cavernous sinus (artery–to–vein) fistulas occur here.

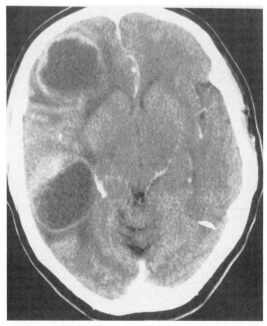

Figure 3.23. Ring-enhancing lesions.

DIFFERENTIAL DIAGNOSIS (DDX) PATTERNS

- "**Ring-enhancing lesion**" DDX: "**MAGIC DR**": **M** = metastases; **A** = aneurysm, abscess; **G** = glioblastoma; **I** = infarct, infectious/inflammatory (toxoplasmosis, lymphoma); **C** = resolving contusion; **D** = demyelinating diseases (multiple sclerosis); **R** = postradiation. (Ring enhancement means the periphery of a lesion gets bright with contrast [Fig. 3.23].)
- **Cerebello-pontine angle (CPA) mass DDX: "AMEN": A** = arachnoid cyst, **M** = meningioma, **E** = epidermoid, **N** = neuroma (schwannoma = acoustic neuroma).
- Grey-white junction lesions DDX: neoplasm (metastases), infection (septic emboli), thrombotic infarcts, vasculitis.
- Skull base meningitis DDX: tuberculosis, fungus, sarcoidosis.
- Periventricular lesions DDX: multiple sclerosis, toxoplasmosis, cytomegalovirus, primary CNS lymphoma, glioma, infarcts, bleeds, age-related changes.
- Basal ganglia density DDX: Hypertensive bleed, TORCH infection, poisoning from lead, carbon monoxide (globus pallidus), or methanol (putamen).

Suggested Readings

Daffner RH. Imaging of vertebral trauma. Rockville, MD: Aspen Publishers, 1988.

Dahnert W. Radiology review manual, 2nd ed. Baltimore: Williams & Wilkins, 1993.

Grossman RI, Yousem DM. Neuroradiology: the requisites. St Louis: Mosby, 1994.

Harris JH Jr, Mirvis S. The radiology of acute cervical spine trauma, 3rd ed. Baltimore: Williams & Wilkins, 1996.

Osborne A. Introduction to cerebral angiography. Hagerstown, MD: Harper & Row, 1980.

Osborn AG. Diagnostic neuroradiology. St Louis: Mosby, 1994.

Ravin CE, Cooper C, Leder RA, eds. Review of radiology, 2nd ed. Philadelphia: WB Saunders, 1994.

Roberge RJ. Facilitating cervical spine radiography in blunt trauma. Emerg Clin North Am 1991;9(4):733–742.

Squire LF, Novelline RA. Fundamentals of radiology, 4th ed. Cambridge: Harvard University Press, 1988.

Weist P, Roth P. Fundamentals of emergency radiology. Philadelphia: WB Saunders, 1996.

Chapter 4

Abdomen/ Gastrointestinal

WHAT TO ORDER AND WHEN

If a computed tomography (CT) or ultrasound (US) is already ordered, abdominal films might not be needed. High-yield indications for abdominal x-rays include: abnormal gas or calcium, free air or perforation, bowel obstruction, ischemic bowel, appendicitis, and toxic megacolon or colitis. Consider a different test for low-yield indications for abdominal x-rays like gallstones (US), fever or abscess (CT), ulcer/epigastric pain (endoscopy or upper GI), GI bleed (CT, nuclear medicine, angiography and/or endoscopy), and chronic pain or palpable mass (CT or US).

- Order barium enema for occult fecal blood, change in bowel habits, or suspicion of colorectal cancer.
- Order upper GI for epigastric pain.
- Order CT for retroperitoneal hematoma; for unexplained blood loss following surgery, femoral artery catheterization, or line placement.
- Order CT with *and without* contrast if the presence of calcium is helpful information (renal or common bile duct stone, abdominal, pelvic, or mediastinal mass) (Figs. 4.1 to 4.4).

X-RAY INTERPRETATION

Remember the "ABCs" of abdominal films: A = air, B = bowel gas pattern, C = calcium, and S = more studies? (see the following).

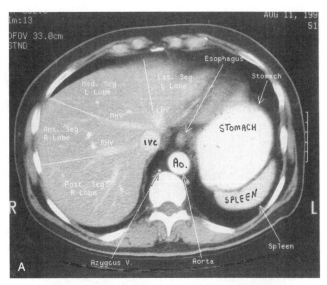

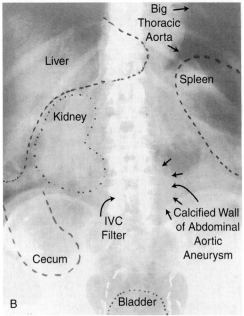

Figure 4.1.

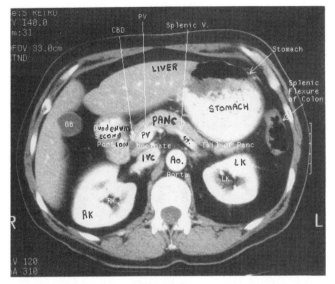

Figure 4.2.

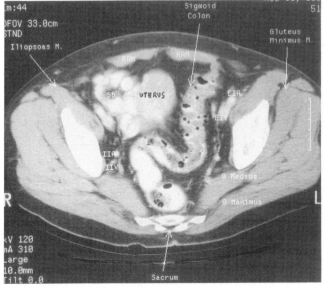

Figure 4.3.

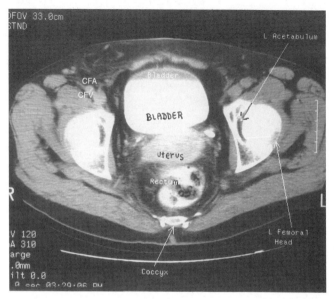

Figure 4.4.

ABDOMEN COOKBOOK CHECKLIST

Free **a**ir?	Chest x-ray better, CT best
Bowel gas pattern?	Dilation, air-fluid levels, obstructed? Wall thick?
Calcium?	Bones & stones
Technique?	Air-fluid levels = upright or decubitus
More **s**tudies?	CT, US, Hepatobiliary scan, barium study

As with the chest x-ray, use a mental checklist and an organized step-by-step approach to dissect the abdominal film with your eyes. Start with technique, noting position and projection, to see if you got what you ordered. Air-fluid levels are the easiest way to check which way is down. If there aren't any, then it is likely a supine film. An AP supine film is also called a "**KUB**" (for kidneys, ureters, and bladder).

Gas outside the bowel lumen is abnormal. An upright or lying on the side (decubitus) film is added to look for **free** (intraperito-

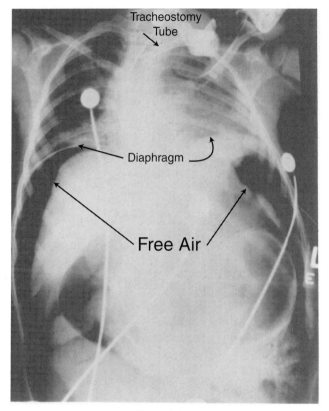

Tracheostomy
Tube

Diaphragm

Free Air

Figure 4.5.

neal) **air** (Fig. 4.5) or bowel **air-fluid levels.** Free air is easier to see on *left* down films, as it outlines the liver edge. Early recognition of free air can be a lifesaver. Do a **quick glance** for black air under the diaphragm, outlining bowel loops, or in the lesser sac or retroperitoneum. Repeat this later. Next, check for **man-made** or foreign objects, tubes, and lines.

Begin the anatomic analysis with the **bones.** Look at the spine pedicles for metastases (the two "eyes" of the AP vertebrae). Check the **liver, spleen, and kidneys** for position, size, and margins. White **calcifications** over the kidneys, ureters, or bladder

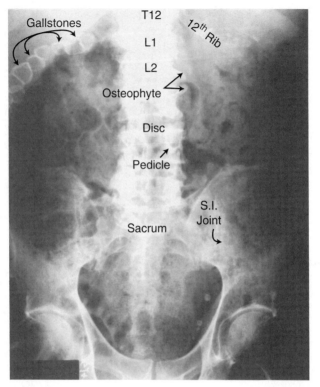

Figure 4.6.

may be kidney stones. Round or laminated calcium in the right upper quadrant may be gallstones (Fig. 4.6). Fifteen to 20 % of gallstones show up on plain film (the inverse of kidney stones, approximately 85%). In a patient with an acute abdomen, layered calcium (concentric rings) in the right lower quadrant that moves when comparing the supine film with the upright film is an **appendicolith,** signifying appendicitis (Fig. 4.7). Mottled or punctate calcium in the epigastric region may indicate a history of chronic pancreatitis (Fig. 4.8).

BOWEL GAS PATTERN

The **bowel gas pattern** is the most difficult part of the film. Imagine yourself as a bubble of gas in a taco being moved by peristalsis

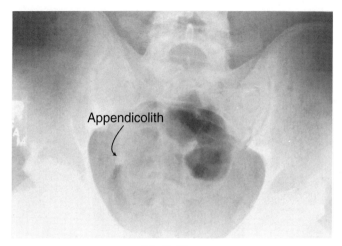

Appendicolith

Figure 4.7.

from stomach to rectum in an intricate plumbing system of squeezing pipes of different sizes. When the plumbing is clogged, the pipes proximal to the blockage get bigger, filling up with black gas and taco from above. The pipes distal to the problem continue to empty and eventually get small and collapse from not being refilled. It's that simple. Try to guess the location of an obstruction, but know that sometimes this will be simply a guess. Gas in a nondilated bowel is usually normal. Small bowel normally has minimal gas, except for infants, and immobile or debilitated patients. Remember the mobile parts of the large bowel with mesentery (sigmoid, cecum, and transverse colon) tend to have gas on supine films and may twist around their mesentery (**volvulus**), causing proximal obstruction. Volvulus in adults occurs in the sigmoid, cecum, and transverse colon, in descending order of frequency. More on midgut volvulus in infants later (see Chapter 6).

OBSTRUCTION VERSUS ILEUS

Big loops of gas-filled large and small bowel seen all over the abdomen in a balanced distribution with long, loose, relaxed air-fluid levels at similar levels may be an **ileus** (adynamic or paralytic). The air-fluid levels in an ileus tend to be long (layer at the

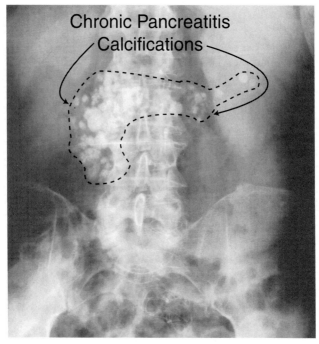

Figure 4.8. Chronic pancreatitis.

same level) because bowel that won't squeeze lets the gas stay flat. The many causes for ileus include postsurgical, posttraumatic, peritonitis, electrolyte imbalance, opiate effects, bowel ischemia, and idiopathic. An ileus may look just like a mechanical large bowel obstruction (or even an early or partial **small bowel obstruction** [SBO]). Serial follow up films may differentiate these and can assess the need for emergent intervention. The distended loops are distributed in a balanced fashion in ileus, but can be more localized in an obstruction. Obstruction also may have more air-fluid levels and more orderly stacked loops. A "**string of pearls**" is multiple small dots of gas in line with each other, but at different levels, in fluid-filled loops of obstructed small bowel. More gas in these obstructed loops results in loops stacked one on top of another in a "**stepladder**" pattern. These configura-

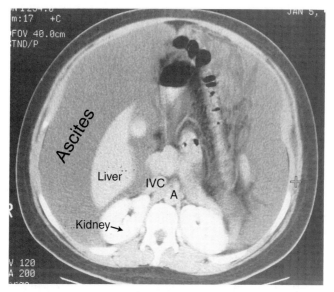

Figure 4.9.

tions result from bowel trying to squeeze and force transit past
an obstruction. Obstruction may be diagnosed with plain radiographs alone; however, barium studies and CT may show the site
and cause.

- **Ascites** causes a diffuse gray haziness on KUB, and floating bowel
  loops in the middle of the belly (the highest point in a patient
  lying down) (Fig. 4.9).
- Bowel gas tends to rise. The rectum is higher when the patient
  is on the stomach (prone). A prone film can differentiate ileus
  from large bowel obstruction because unblocked gas is allowed
  to rise to the rectum in an ileus.

OBSTRUCTION

An **obstruction** may have focally dilated bowel with air-fluid levels
that are shorter in length, have tight turns, and may be at many
different heights (Fig. 4.10). **Causes of obstruction** include **adhesions** (50% and the number one cause of SBO), **hernia** (15%

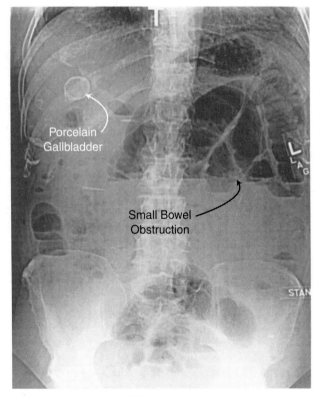

Porcelain
Gallbladder

Small Bowel
Obstruction

Figure 4.10.

and the number one cause of SBO in those without a surgical history), **volvulus,** and **tumor.** Complications of obstruction can sometimes be seen on CT scan. These surgical emergencies include "strangulation" (ischemia with mesenteric vascular compromise), "closed-loop obstruction" (incarceration or blockage on both sides of a loop), and free air. These are surgical emergencies. **Bowel wall thickening** may be an early warning sign of bowel at risk for dying (infarction) or popping (perforation).

- Bowel gas over the obturator foramen with an SBO may indicate inguinal hernia.
- In infants, large and small bowel look the same; thus, it is hard

to tell the type of obstruction. Make a guess based on location and number of loops.

BOWEL WALL THICKENING

Abnormally thick bowel walls manifest as folds too thick, or loops too far apart from each other. Fold thickening may occur in gastric rugae, duodenal folds, small bowel valvulae conniventes (plicae circulares), or colonic haustra. Small bowel folds as thick as a pen are too thick. Thick haustrae are called "thumb printing," as the thick loop separation looks like someone stuck their thumb on the haustra between the loops (Fig. 4.11). **Causes of bowel wall thickening include blood, pus** (colitis or enteritis), **water** (edema or hypoproteinemia), **cells** (carcinoma, lymphoma, or metastatic tumor infiltration), **ischemia** (volvulus, obstruction, or clot), **or** any **inflammation** (duodenitis, gastritis). A loop of focally dilated or inflamed bowel may point to local inflammation adjacent to pancreatitis, appendicitis, or cholecystitis, and is called a "sentinel loop."

• A "picket fence" or "stack of coins" appearance of the small bowel may indicate bleeding within the bowel wall.

FREE INTRAPERITONEAL AIR

An untreated obstructed or ischemic bowel may lead to **free** (intraperitoneal) **air** or air in the bowel wall (**pneumatosis** intestinalis), indicating perforation or infarction. The supine abdominal film is not sensitive for free air. Free intraperitoneal air may be seen on upright or decubitus abdominal films, but the upright chest x-ray is the most sensitive plain film, and should be included in the acute abdomen series. Fifteen percent of emergency patients with abdominal complaints have chest abnormalities, and lower lobe pneumonia may present with belly pain. Free air appears as a dark linear band of air under the diaphragm (Fig. 4.5), or outside the liver on a left down decubitus film. On a supine film, free air may outline the falciform ligament and forms a dense line from the umbilicus to the right upper quadrant, more commonly seen in infants. Free air may outline both sides of the bowel wall (Rigler's sign). Pneumatosis appears as linear or cystic lucencies overlying the bowel wall. Retroperitoneal air may form a dark outline around the kidney or psoas muscle. Don't miss these ominous findings.

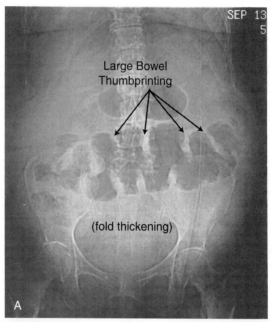

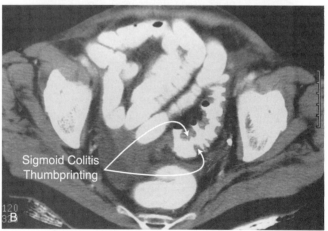

Figure 4.11.

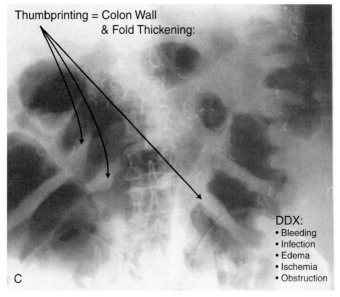

Thumbprinting = Colon Wall
& Fold Thickening:

DDX:
• Bleeding
• Infection
• Edema
• Ischemia
• Obstruction

C

Figure 4.11. *(continued)*

• Free intraperitoneal air is normal postoperatively, and usually resolves in a week to 10 days, but may take up to a month.
• Supine films are not sensitive for free air.
• On CT, free air rises to the top (anterior) (Fig. 4.12).
• With trauma or an acute abdomen, flip through abdominal CT with lung windows to exclude free air.

VOLVULUS

Sigmoid volvulus is a dilated, upside-down, U-shaped closed loop, twisted upon its mesentery. It shows up in the mid abdomen, often shaped like a coffee bean, tapering towards the pelvis, with a vertical line in the middle of the upside-down "U" representing the walls of adjacent loops of sigmoid touching each other (Fig. 4.13). A large bowel obstruction proximal to a sigmoid volvulus is the rule, unlike the less common **cecal volvulus** which can cause a small bowel obstruction. Cecal volvulus typically shows up as a big oval loop in the midabdomen or left upper quadrant (Fig.

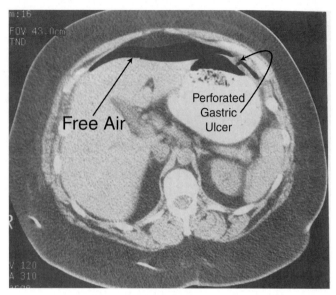

Free Air

Perforated Gastric Ulcer

Figure 4.12.

4.14). The big loop twists and flips into these nonanatomic locations. Sigmoid and cecal volvulus may be hard to differentiate without a contrast enema or sigmoidoscopy. An obstruction causes proximal bowel to dilate; therefore, any large bowel obstruction can also lead to a secondary small bowel obstruction.

INTUSSUSCEPTION

Intussusception is telescoping of the bowel. It is much more common in children where it is usually idiopathic. Most adults with intussusception have a pathologic mass as a lead point causing the bowel to telescope upon itself.

TOXIC MEGACOLON

Toxic megacolon is an acute dilation of the colon in a patient with systemic signs and an underlying illness like ulcerative colitis (inflammatory bowel disease), pseudomembranous colitis (antibiotic-associated colitis from *Clostridium difficile*), or ischemic

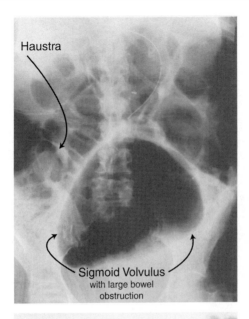

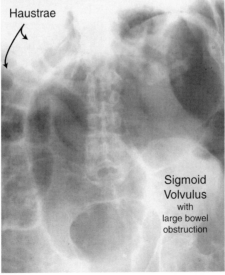

Figure 4.13.

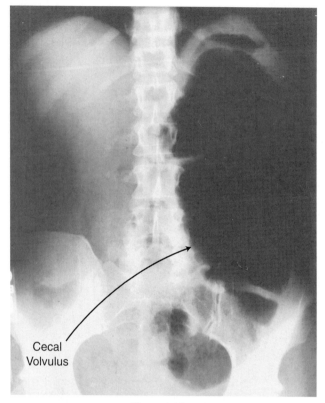

Cecal
Volvulus

Figure 4.14.

bowel. Due to the risk of perforation, contrast enema is contraindicated in toxic megacolon.

- The transverse colon in toxic megacolon averages 9 cm in diameter, but may get bigger (greater than 6 cm is sometimes abnormal).
- The cecum tends to have a larger diameter than the remainder of the colon (greater than 9 cm is often abnormal) and may be the first part to perforate (La Place's law: more wall tension with greater radius).

GI BLEED

With an acute **GI bleed,** consider endoscopy, arteriography
($+/-$ embolization), nuclear medicine bleeding study, vasopressin infusion, or surgery. GI bleeding must be fast (greater
than 0.5 to 1 cc/min) and currently active to see on angiography.
Nuclear medicine bleeding scan is much more sensitive for active
and recent bleeding (greater than 0.1 cc/min). Another advan-
tage of the nuclear medicine scan is that the patient with intermittent bleeding may be injected with the radiopharmaceutical and
then scanned later, whenever bleeding recurs (to find the
source). The intermittent bleeder is harder to catch with arteriography, but a positive study is more definitive, and embolization
may be considered. The sooner imaging occurs after active bleeding, the more likely the source will be discovered. Thus, don't
delay in calling interventional radiology or nuclear medicine immediately.

    Mesenteric ischemia (intestinal angina) is a difficult diagno-
sis, often occurring in patients with atherosclerotic disease. A
clue may be bowel wall thickening, typically near the "watershed
zone" of the splenic flexure, between superior and inferior mesenteric arterial supplies. Definitive diagnosis requires angiography.

    In 90% of patients with a GI bleed, the cause is proximal to
the ligament of Treitz. In most of these, the nasogastric tube
aspirate is bloody. Thus, an upper GI source must be excluded
prior to attributing melena or bloody stools to a more distal problem. While thinking, order an upright film, give fluids or blood
products, check labs, and lavage the stomach. If a decubitus film
is done, it should be *left* side down to allow any free air in the
lesser sac to enter the big peritoneal cavity and outline the liver
edge (as with an ulcer perforation into the lesser sac).

BARIUM STUDIES/FLUOROSCOPY

Barium studies look at structure and function by showing mucosal
ulcerations, polyps, or masses, while also looking at motility,
transit, and causes for obstruction. Adhesions may be seen as
fixed, tethered, spiculated loops of small bowel. A colon tumor
may be a dark defect in the pool of white barium, or may look
like an "apple core" if it surrounds and narrows a loop of colon
(normally apple-sized).

The screening enema is a cheap way to check the colon out in older patients. Heme-positive stools require looking for a source. Most colorectal cancers grow from adenomas or adenomatous polyps, and the sigmoid colon is the most common location.

- The chance of cancer in colonic polyps (adenomas) depends on the size and shape of the polyp: less than 1 cm = 1% risk, 1 to 2 cm = 10% risk, greater than 2 cm = 25 to 50 % risk. Sessile or villous is more risky than tubular or pedunculated.
- Only breast and lung cancer cause more cancer deaths than colon cancer.

BARIUM ORDERS/TECHNIQUE

- Upper GI, small bowel follow through, and barium enema contrast is much more dense than CT oral contrast, so the CT should be done first if both are going to be done (to avoid CT artifacts).
- Water-soluble gut contrast agents (like concentrated gastrografin) are hypertonic and can cause hypovolemia and increased bowel distention.
- If there is recent or upcoming bowel surgery or concern for perforation, then use water soluble oral or rectal contrast for CT or fluoroscopic GI studies. Spilled barium can cause peritonitis. Spilled water-soluble contrast gets easily resorbed.
- Gastrografin is bad to aspirate (causes pneumonitis). In elderly or aspiration prone patients, barium or nonionic contrast is better.
- **Prep** for an ''upper GI'' or ''small bowel follow through'' barium study requires nothing by mouth after midnight, plus an nasogastric tube if there is concern for obstruction. (An upper GI ends at the ligament of Treitz.)
- **Prep** for a contrast enema may include 2 days of a liquid diet, a night-before cathartic, and an early morning suppository. Prepare the patient psychologically for this study.
- Esophago-gastro-duodenoscopy (EGD) and colonoscopy are performed by the gastroenterologists, and may be alternatives to barium studies in some instances. Sigmoidoscopy only looks at the distal colon and is also done by some surgeons.

CT ABDOMEN

With new scanners, **triple contrast** CT (oral, rectal, and intravenous) is sensitive for large masses, appendicitis, diverticulitis, and abscess. Abscess search is a CT emergency, and abscess drainage catheters can be placed under CT guidance. CT scanning during intravenous injection of contrast provides dynamic information about vascularity, which may be especially useful with liver lesion characterization.

Benign and malignant **liver lesions** may look identical unless a dynamic CT is done with contrast, which can usually differentiate these. Dynamic implies the same area is viewed during different vascular phases (i.e., arterial [early] or venous [later] phase). For example, benign liver hemangiomas initially appear as dark lesions on CT, but eventually fill in from the periphery with contrast in a nodular fashion. Nuclear medicine liver spleen scan can also diagnose liver hemangiomas larger than 2 cm, but MR plays the central role in the evaluation of indeterminate liver lesions.

- A pelvic CT may or may not be included in an abdominal study. Dependent-free fluid or blood may only be seen in the pelvis.
- Remember portal triad relationships with Mickey Mouse's face as an axial image; the head is the portal vein, the right ear is the common bile duct, and the left ear is the hepatic artery (the ears are more anterior, like the top of a CT).
- **Normal sizes:** spleen less than 12 cm, gallbladder less than 5 cm, small bowel diameter less than 3 cm, cecum less than 9 cm, noncecum colon less than 6 cm, common bile duct less than 7 mm or less than the number of decades in age (in mm).
- Dependent bright blood cells layering horizontally in a blood collection with darker plasma above is known as the "hematocrit effect" on CT or US.

GALLSTONES/CHOLECYSTITIS

Gallstones that get inflamed or obstruct the cystic duct can cause cholecystitis, or an inflamed biliary system. US and nuclear medicine cholescintigraphy (**hepatobiliary scan**) have similar sensitivity rates for **acute cholecystitis.** With acute right upper quadrant colic, US is usually the initial screening exam, and is usually more available. Hepatobiliary scan better tells *if* obstructed, whereas US better tells *why* obstructed. **US is the way to diagnose gallstones and gallbladder morphology, whereas hepatobiliary scans show cystic duct obstruction** (which is the underlying event in acute cholecystitis). One approach is to use hepatobiliary scans when the US is equivocal with known gallstones, or if the clinical suspicion for acute cholecystitis is high. CT is useful for atypical symptoms, where a broader differential diagnosis is being considered.

US shows bright (echogenic) stones with dark bands (shadows) behind them. Gallbladder inflammation will cause wall thickening and fluid (dark) around the gallbladder or in its wall (Fig. 4.15). If there is common bile duct blockage, the duct may dilate. Distal bile duct stones can also cause pancreatitis, resulting in a big, dark pancreas on US.

In a **hepatobiliary scan,** an IV radiopharmaceutical is excreted into the biliary tract. The main feature of the positive hepatobiliary scan is the absence of gallbladder filling, implying a blocked cystic duct (Fig. 4.16). Common bile duct obstruction causes nonvisualization (nonfilling) of the common bile duct, duodenum, and gallbladder. The hepatobiliary scan can also give a gallbladder ejection fraction that tells how well the gallbladder contracts. The ejection fraction may be low in chronically hospitalized or ICU patients with *chronic* cholecystitis or acalculous cholecystitis (without stones).

Prep for a hepatobiliary scan includes between 4 to 24 hours of fasting; the false positive rate goes up if less than 4 hours or greater than 24 hours.

PANCREATITIS

Pancreatitis is best evaluated with a dynamic CT scan (or US) that can see a swollen, edematous, or hemorrhagic pancreas (Fig. 4.17), pancreatic duct dilatation, stones, and the complications of pancreatitis, like pseudocyst, phlegmon, abscess, necrosis, and pleural or peritoneal effusions. Imaging in acute pancreatitis is helpful with persistent pain, leukocytosis, recurrent fever, or suspicion of necrosis or infection which could require surgical debridement. Lesser sac pseudocyst may need drainage through the skin (by CT or US guidance) or stomach (by endoscopy), and abscess needs surgical or percutaneous drainage (see Chapter 8).

To better assess the biliary and pancreatic ductal systems, gastroenterologists do endoscopic retrograde cholangiopancreatography (**ERCP**), whereas radiologists approach the intrahepatic biliary system through the skin with fluoroscopic guidance. ERCP is duodenoscopy with cannulation of the ampulla of Vater, and injection of contrast into the biliary and pancreatic ducts under direct visualization and fluoroscopic guidance. Baskets can

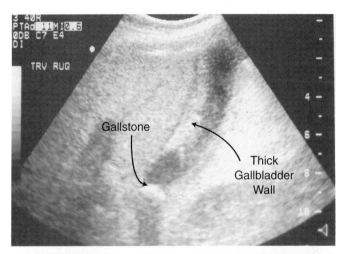

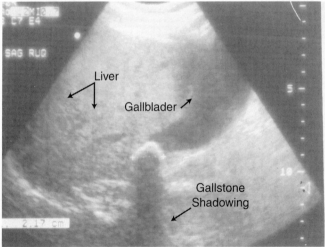

Figure 4.15. Ultrasound.

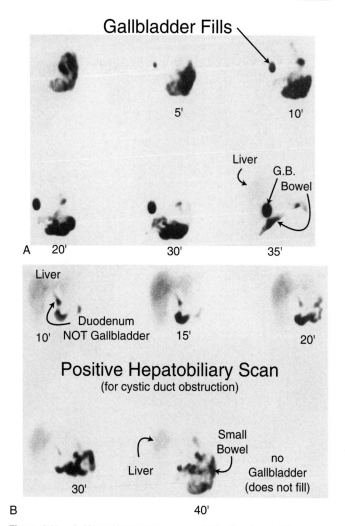

Figure 4.16.   A. Normal hepatobiliary scan. B. Positive hepatobiliary scan.

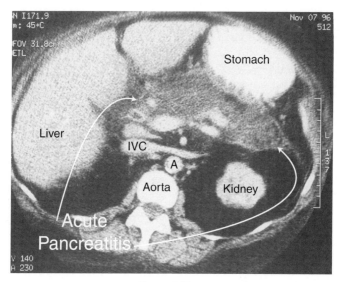

Figure 4.17.

be used for scooping out stones; narrow strictures or blockages can be opened with balloons and metal or plastic stents.

- Postpancreatitis liquification (fluid) is called a "pseudocyst," whereas a "phlegmon" is partly solid.
- Pancreatic calcifications usually indicate chronic pancreatitis due to alcohol (Fig. 4.8). Thus, consider other alcohol issues when seen (withdrawal, DTs, seizure).
- A developmental anomaly called pancreas divisum predisposes to pancreatitis. In a divisum, the minor duct (Santorini) drains all but the uncinate process, which is drained by the abnormally short major duct (Wirsung), which normally drains most of the pancreas.

APPENDICITIS

Appendicitis is usually a clinical diagnosis. In equivocal cases, plain radiography, US, or CT may come in handy. A calcified **appendicolith** in the right lower quadrant confirms the diagnosis, but is seen in less than 30% of appendicitis CTs, and even fewer plain films (Fig. 4.7). CT or US may help identify fluid collections,

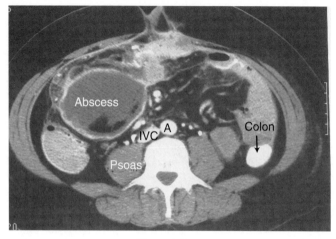

Figure 4.18.

pericecal abscess, phlegmon, or a big, fluid-filled appendix. US avoids radiation to young or pregnant patients and has a high positive predictive value. However, CT is more definitive and sensitive than US, is less operator-dependent, and may also show adjacent inflamed fat. Normal fat is dark on CT. Fat with hazy, streaky brightness next to the appendix is inflamed. CT can decrease the negative laparotomy rate.

Unruptured appendicitis on US is a noncompressible, dilated, thick-walled tube, without peristalsis. Rupture appendicitis may wall itself off, forming an abscess full of dark fluid or bright pus. Too much gas in the abdomen dampens the sound waves and can make the US uninterpretable.

DIVERTICULITIS/ABSCESS

Diverticulitis is commonly located in the sigmoid colon and is depicted on CT as pericolic inflammatory infiltration (bright) in the fat adjacent to diverticular outpouchings. CT can unearth occult abscess as a dark fluid collection with a bright thick rim (Fig. 4.18). The presence of gas within a fluid collection that is definitely not bowel is suspicious for **abscess**. Fluid can be hard to tell apart from bowel. Fluid or abscess usually wrap around loops of bowel and appear amorphous. Diagnosis and treatment of diverticular abscess can often be made with CT-guided drainage catheter placement, which can make surgery easier or unnecessary.

INFLAMMATORY BOWEL DISEASE

Inflammatory bowel disease can be evaluated with contrast studies and/or CT. **Crohn disease** most commonly affects the terminal ileum and may have fistulas, skip areas, "cobblestone" mucosa, or string-like strictures. **Ulcerative colitis** may show bowel wall thickening, ulcerations, toxic megacolon, and continuous involvement from the rectum proximal. Crohn disease has deeper, transmural involvement.

• #1 cause for enterovesical (small bowel to bladder) fistulas is Crohn disease.
• #1 cause of colovesical (colon to bladder) fistulas is diverticulitis.
• The much more common cause of gas in the bladder on CT, however, is just a normal bladder catheter.

ABDOMINAL TRAUMA

Abdominal CT competes with the faster, less specific diagnostic peritoneal lavage (DPL) in screening for serious visceral injury or the complications of **abdominal trauma** (controversial). On CT, blood around the liver, spleen, kidney, or in the paracolic gutters, hepatorenal pouch of Morison, lesser sac, mesentery, or the pelvic pouch of Douglas may indicate significant trauma. Fluid should be measured with **CT numbers** to see if it is fresh blood density. Recent bleeding is usually, but not always, bright. Likewise, if the abdomen is full of blood, the fluid that is slightly brighter should be near the organ that is bleeding. The size of the inferior vena cava (IVC) mirrors blood volume status. Enhancement of a thick bowel wall with a small IVC (relative to the aorta) suggests "shock bowel."

Liver, spleen, or kidney **laceration** may look like a dark, jagged, serrated band shaped like a lightning-bolt, or may be bright if actively bleeding (Figs. 4.19 and 4.20). Liver laceration may be complicated by biloma or pseudoaneurysm. Serial CTs may screen for delayed splenic rupture, which may take days to manifest. CT can differentiate surgical emergencies from conservatively treated trauma injuries. Some trauma centers routinely use

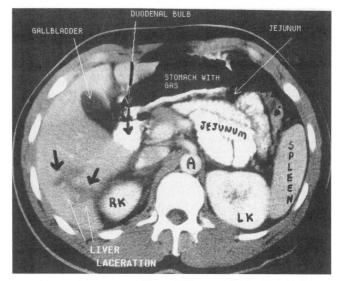

Figure 4.19.

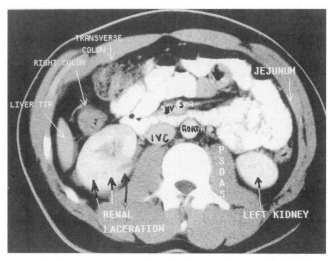

Figure 4.20.

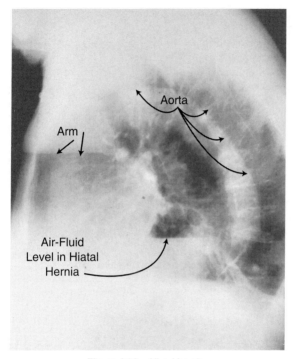

Figure 4.21.   Hiatal hernia.

US in the emergency room to look for free fluid in the abdomen and pelvis as a secondary sign of organ injury.

SUMMARY

After completion of the anatomic analysis, remember to check your "psychological blind spots," including the lung bases, costo-phrenic angles (CPAs), soft tissues, and any personal problem areas. Look for free air again. Are follow up films needed? Would an additional CT, US, or barium study change management or prognosis?

PEARLS WITH NAMES

• **Klatskin** tumor = cholangiocarcinoma at hepatic hilum/bifurcation of intrahepatic bile duct.

- **Virchow**'s node is supraclavicular adenopathy from a gastric cancer.
- **Boerhaave** syndrome is complete perforation of the lower esophagus wall with spillage of the stomach contents, blood, and gas into the mediastinum or pleural space. Presents in trauma patients with left effusion or hydropneumothorax, but without hematemesis (versus Mallory-Weiss).
- **Mallory-Weiss** tears (no perforation) are mucosal/submucosal, longitudinal, and occur near or just below the gastroesophageal junction in patients with repeated forceful vomiting and hematemesis (classically in alcoholics).
- Diaphragmatic hernias: **Bochdalek** = **B**abies (younger), **B**igger, **B**ack (posterior), left, versus **Morgagni** = **M**ature (older), **M**inuscule, **M**edial (right/anterior); Hiatal (Fig. 4.21).

Suggested Readings

Balthazar EJ, ed. The radiologic clinics of North America: imaging the acute abdomen. Philadelphia: WB Saunders, 1994:32(5).

Brewer BJ, Golden GT, Hitch DC, et al. Abdominal pain. An analysis of 1000 consecutive cases in a university hospital emergency room. Am J Surg 1976;131:219.

Chen MYM, Pope TL, Ott DJ, eds. Basic radiology. New York: McGraw-Hill, 1996.

Dahnert W. Radiology review manual, 2nd ed. Baltimore: Williams & Wilkins, 1993.

Halpert RD, Goodman P. Gastrointestinal radiology: the requisites. St Louis: Mosby Year Book, 1993.

Laufer I, Levine MS. Double contrast gastrointestinal radiology, 2nd ed. Philadelphia: WB Saunders, 1992.

Mirvis SE, Shanmuganathan K. Abdominal computed tomography in blunt trauma. Semin Roentgenol 1992;27(3):150–183.

Ravin CE, Cooper C, Leder RA, eds. Review of radiology, 2nd ed. Philadelphia: WB Saunders, 1994.

Rumack CM, Wilson SR, Charboneau JW. Diagnostic ultrasound. St Louis: Mosby Year Book, 1991.

Squire LF, Novelline RA. Fundamentals of radiology, 4th ed. Cambridge: Harvard University Press, 1988.

Webb WR, Brant WE, Helms CA. Fundamentals of body CT. Philadelphia: WB Saunders, 1991.

Weist P, Roth P. Fundamentals of emergency radiology. Philadelphia: WB Saunders, 1996.

Weltman DI, Zeman RK. Acute diseases of the gallbladder and biliary ducts. Radiol Clin North Am 1994;32:933–950.

Chapter 5

Genitourinary

The genitourinary (GU) system is most often evaluated in the emergency setting with a plain abdominal radiograph, ultrasound (US), intravenous pyelogram (IVP), or computed tomography (CT). Vaginal bleeding, pelvic pain, hematuria, and scrotal pain are a few of the emergent scenarios with which to become comfortable. The first step in GU is figuring out when and what study to order.

PLAIN FILM (KUB)/HEMATURIA/CALCIFICATION

The **KUB** abdominal film is approached in the same methodical manner, regardless of the reason for the film (see chapter 4). In the setting of **hematuria,** make a close search for calcium along the course of the kidneys, ureters, and bladder. **Unenhanced CT,** however, will pick up more small stones than a KUB (Fig. 5.1). **Calcifications** in the pelvis could be stones in the ureterovesicular junction, or simply harmless phleboliths (venous calcium). Vascular calcifications are curvilinear or concentric, but may be confused for stones if they overlie the region of the distal ureter. IVP or CT may be necessary to sort this out; a rim of soft tissue around the calcium on CT suggests distal ureteral origin. In the pelvis, popcorn calcifications are likely uterine fibroids (Fig. 5.2), and teeth-like calcifications may signify a dermoid or teratoma.

Differential diagnosis (DDX) for medullary **nephrocalcinosis** (diffusely calcified renal medulla): hyperparathyroidism, renal tubular acidosis, medullary sponge kidney, and hypercalcemic states.

ULTRASOUND / NUCLEAR MEDICINE

Ultrasound can screen for hydronephrosis. Kidney and bladder calculi show up on US as bright foci with shadowing behind.

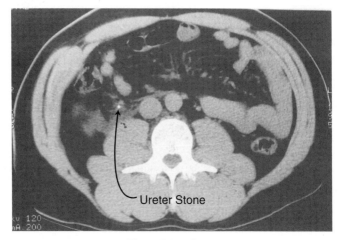

Figure 5.1. Stone.

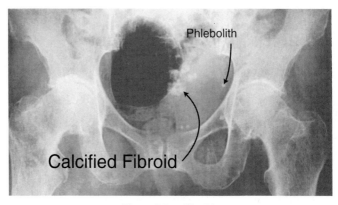

Figure 5.2. Fibroid.

Ureter stones and nondilated ureters, however, are difficult to impossible to see with US. IVP and **nuclear medicine** renal scans can better evaluate function. The nuclear medicine scan also looks at blood flow, which is helpful with renal transplants and renal artery stenosis.

UNENHANCED CT

If the clinical diagnosis is uncertain in the setting of acute flank pain, **unenhanced CT** is an appropriate alternative to IVP. Thin slices using helical technique may be necessary to look for small ureter stones (Fig. 5.1). CT provides valuable information about processes outside the urinary system.

MISCELLANEOUS PEARLS

- Gas in the bladder that has not been instrumented or catheterized must be presumed to be due to bowel to bladder fistula.
- **Emphysematous pyelonephritis** (and emphysematous cholecystitis) are necrotizing infections from E. coli in diabetics and have gas in the wall or lumen.
- Tuberculosis causes **sterile pyuria** (white cells in culture-negative urine) and affects the upper urinary tract before the lower tract.

IVP

The **IVP** is a cheap and readily-available screening study for patients with suspected urinary tract pathology, flank pain, or hematuria. The IVP follows the path of intravenously administered contrast (and urine) from kidney to toilet. For 15 to 20 minutes following IV contrast bolus, sequential x-rays are taken of contrast in the kidneys, ureters, and bladder, looking for **dilation, blockage, stones, or masses** (Figs. 5.3 and 5.4). Hematuria is the most common indication, which could be from infection, tumor, trauma, or other inflammatory process. The IVP causes diuresis, which may dislodge a stuck stone, relieving the obstruction. Generally, stones larger than 3 mm in the ureter won't spontaneously pass. In this case, aggressive hydration, ureteral stenting, lithotripsy, ureteroscopic retrieval, or percutaneous removal by urology with interventional radiology may help avoid open surgery (Figs. 5.5 and 5.6).

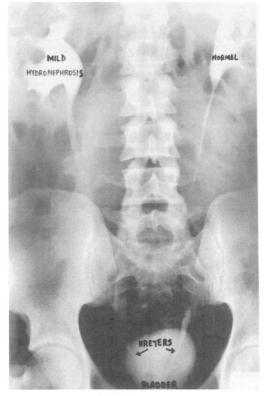

Figure 5.3. IVP.

A dark spot or "**filling defect**" in the white IVP contrast could be a stone, blood clot, tumor, fungus ball, or sloughed papilla in papillary necrosis. **Kidney stones** are dense on x-ray approximately 85 to 90% of the time (inverse of gallstones), with the most common salt being calcium oxalate. The uncommon urinary stones that aren't radiodense include uric acid (lucent) and cystine (mildly opaque). The most common place for stones to lodge is the ureterovesicular junction, where the ureters tunnel in the posterolateral aspect of the bladder base. Renal masses may distort the renal contour or displace collecting system architecture.

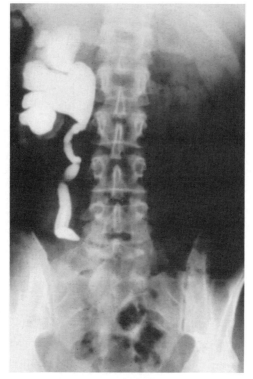

Figure 5.4.   Marked hydronephrosis and hydroureter on IVP.

IVP PEARLS

- The most common salt in **staghorn calculi** (a stone that fills the renal pelvis) is struvite (magnesium ammonium phosphate—often laminated) from urea-splitting bacteria like proteus (Fig 5.6).
- **Uric acid stones are the most common radiolucent stones** and are associated with concentrated acidic urine, hyperuricosuria, gout, myeloproliferative disorders, and diuretics like furosemide.
- Stasis, poor emptying, and gram-negative urinary tract infections predispose to bladder calculi.

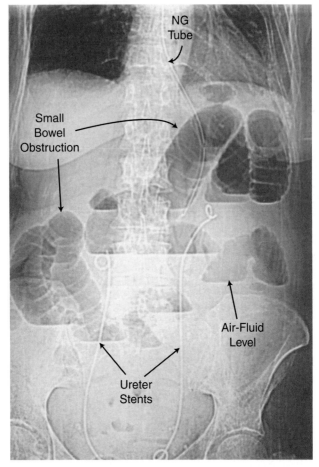

Figure 5.5. Upright KUB with small bowel obstruction.

• DDX for **papillary necrosis** = ''SODA'': S = sickle cell anemia, O = obstruction, D = diabetes, A = analgesics.

MASSES AND CYSTS ON US, CT, OR MR

The ureters can be displaced by lymphadenopathy, **masses,** aortic aneurysms, or retroperitoneal processes. Often, a question raised

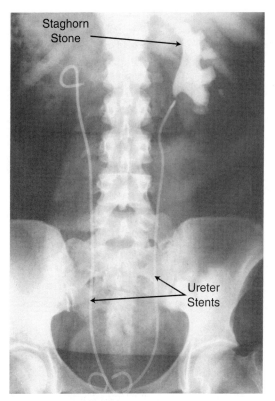

Staghorn
Stone

Ureter
Stents

Figure 5.6. Staghorn stone.

on IVP will need to be addressed or better characterized by **CT or US.** This includes masses, cysts, ureter narrowing, and contour irregularities. Renal cysts are best evaluated first by US. **Cysts** are "**simple**" if there are thin walls, no internal echoes, and brighter echoes behind the cyst (increased "through-transmission"). Cysts are "**complex**" if there are echoes (debris, blood, pus, necrosis, etc.). Solid renal masses are best evaluated by CT (or MR). Water density cysts have CT numbers 0 to 20. If there is fat within the mass, CT is diagnostic of benign angiomyolipoma (negative CT numbers identify fat).

If a CT scan is done to clarify a question of stones, an unenhanced CT may suffice. For the question of renal masses however, CT should be ordered both with *and* without contrast to assess for enhancement of solid lesions or masses (see Fig. 10.5).

In the kidney, dynamic MR may offer a simultaneous look at structure and function in the setting of renal neoplasms, cysts, renal artery stenosis, or unexplained urine cytology atypia. This could someday substitute for CT, US, and nuclear medicine renal scans in some cases, but is not yet proven, and may not yet be available at your institution.

- Helical CT allows for three-dimensional reconstructions and CT angiography studies, which may be useful with renal artery stenosis, ureteral stones, and renal artery relation to aortic aneurysm (see Fig. 11.8).
- Transitional cell carcinoma is the most common ureter or bladder tumor and tends to be multifocal. Ninety percent occur in the bladder (polypoid is more common than infiltrating).
- Renal cell carcinoma tends to invade the renal veins and IVC, which affects staging. (Adrenal and hepatocellular carcinomas also invade the IVC.) (See Fig. 10.5.)
- Retroperitoneal fibrosis may be mass-like and occurs with tumors, aortic aneurysms, and certain drugs.

TRAUMA

The male patient with pelvic **trauma** requires a retrograde urethrogram prior to bladder catheterization. If necessary, this can be done in the emergency department with portable films. If a bladder catheter has already been inadvertently placed, do a urethrogram around the catheter; do not remove the catheter first!

If the patient is too unstable to wait for CT or IVP, a one-film IVP on the way to surgery can be performed in the emergency department. This can confirm the presence of two kidneys, but can't reliably exclude major renal injury. Unilateral absence of excretion, however, confirms major renal injury.

If there is time, **CT** best evaluates for renal **lacerations,** fractures, subcapsular hematomas, and other traumatic injuries (see Fig. 4.20). CT can differentiate surgical emergencies from conservatively treated injuries in the kidney. **Minor** (contusion or laceration) **and major injuries** (extension to renal collecting system) are usually observed, whereas **catastrophic or critical** injuries (fragmentation $+ / -$ vascular pedicle injury) require emergent surgery.

TRAUMA PEARLS

The pouch of Douglas or recto-uterine cul-de-sac is a good place to look for intraperitoneal fluid or blood.

CT can qualify **bladder ruptures,** as well as the commonly associated pelvic fractures. Extraperitoneal rupture is more common than intraperitoneal, and may be managed conservatively with bladder rest and a suprapubic catheter. Intraperitoneal rupture is usually from blunt trauma to a full bladder and requires surgery. **CT cystogram** is performed by filling the bladder with diluted contrast via a bladder catheter (after clearing the urethra clinically or with a urethrogram). Gross hematuria is the most reliable sign of urinary tract injury; however, in the setting of trauma, microscopic hematuria alone does not require further evaluation.

TESTICLE

The **testicle** is well evaluated with US (with Doppler to look at blood flow). The ''bell-clapper'' deformity allows the testicle more motion and predisposes to **testicle torsion** (twisting that cuts off the blood supply (see Chapter 6). Varicocele may cause infertility and can be surgically ligated. Embolization by interventional radiology is usually reserved for recurrence. Testicular malignancies are hypoechoic or heterogeneous on US. The most common type of primary testicular neoplasm is **seminoma,** which is hypoechoic, well-defined, and has the best prognosis.

PELVIC ULTRASOUND

Abnormal vaginal bleeding, pelvic pain, and adnexal masses can be assessed with transabdominal and transvaginal US. Transabdominal exam requires a full bladder as a window into the pelvis. Transvaginal requires an empty bladder. Pelvic US provides a quick look at the endometrium, cervix, adnexa, ovaries, bladder, cul-de-sac, and any loops of bowel that happen to be fluid-filled. Remember that **ultrasound can't see through gas.**

A common scenario is pelvic or adnexal pain with a tender complex **adnexal mass,** with or without vaginal bleeding

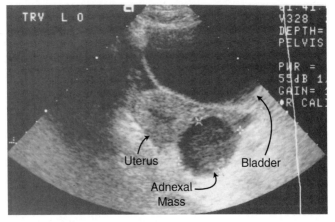

Figure 5.7. Pelvic ultrasound showing adnexal mass.

(Fig. 5.7). The differential diagnosis in this case is broad and includes **ovarian neoplasm, metastasis, bleeding ovarian cyst, ectopic pregnancy, tubo-ovarian abscess, ovarian torsion, and endometrioma.** The history will often be the most important clue in this setting. Complex means mixed echogenicity, thus not simply cystic and not completely solid either, possibly with debris, septations, loculations, bleeding, or necrosis. Simple ovarian cysts can bleed, rupture, or twist, thus becoming complex. **Dermoids** can have hair, teeth, or fat giving a fluid-fluid level. **Torsion** (twisting) of the ovary or of an adnexal mass cuts off the blood supply and might show decreased or absent blood flow on Doppler. It is more common in childhood and adolescence.

Ectopic pregnancy has a positive B-HCG, many US appearances, and may even have the classic triad of **pelvic pain, vaginal bleeding, and a palpable adnexal mass** (most do not). Pain is almost always present, whereas the palpable mass is much less common. **Normal pregnancies are seen by 5 weeks** on transvaginal US. Beware of the ''pseudogestational sac'' of ectopic pregnancy, where hormonal stimulation of the uterine lining simulates a normal gestational sac. A normal sac should burrow its way next to the uterine cavity, creating a ''double decidual sac sign.''

Seeing an intrauterine pregnancy virtually excludes an ectopic pregnancy (exceptions occur in 1 in 7000 to 30,000 pregnancies); however, up to 1% of infertility patients on pro-ovulatory medications will have simultaneous ectopic and intrauterine pregnancies. Ectopic pregnancy may show a complex adnexal mass sometimes with a tubal ring around a hypoechoic (dark) center. Sometimes the only finding may be hypoechoic (dark) or complex (mixed echo) fluid in the cul-de-sac, with some bright echoes in this fluid from clotted blood or hematosalpinx. Color Doppler may show a "ring of fire" with high velocity, low impedance flow. Ectopic pregnancy occurs in 1.4% of all pregnancies and usually shows itself by 7 weeks menstrual age. The fallopian tubes are the location in over 95% (ampulla or isthmus).

Ultrasound may also be helpful with a variety of **gynecological and obstetrical emergencies. Pelvic inflammatory disease** (PID) may show an enlarged endometrium from endometritis, a fluid or pus-filled fallopian tube in hydrosalpinx or pyosalpinx, or a complex multilocular septated adnexal or cul-de-sac mass with debris in a **tubo-ovarian abscess.**

The pregnant patient with bleeding or pain may be screened for spontaneous abortion, subchorionic hematoma, or placental abruption (blood under, or more commonly, at the edge of placenta). If there is prolonged bleeding or infection after an elective or spontaneous abortion, endometrial wall thickness greater than 5 mm suggests retained products of conception. US may help evaluate the **1 in 4 pregnancies** that **have significant first trimester bleeding (of which** ½ **spontaneously abort).** US also may diagnose fetal demise, intrauterine growth retardation, fetal hydrops, gestational trophoblastic disease, and a variety of fetal anomalies, malformations, and **congenital defects** (such as heart disease, neural tube defects, kidney obstruction, and anencephaly).

• **B-HCG** should more than double every 2 days early on. B-HCG that drops slowly suggests ectopic pregnancy or retained products of conception. A slow rise may also signify an ectopic pregnancy.

• B-HCG levels are highly lab-specific, depending on which standard is used.

MR AND WOMEN'S IMAGING

MR may be useful to further evaluate ambiguous or atypical causes of pelvic masses or pain. MR is the study of choice in the evaluation of **endometriosis** (abnormal growth of endometrial tissue outside the uterus) or **adenomyosis** (abnormal ingrowth of endometrial glands into the myometrium). MR accurately stages many pelvic malignancies. Postmenopausal bleeding in a patient without hormonal replacement is endometrial cancer until proven otherwise. US shows endometrial thickening well. MR or US hysterography may be diagnostic of uterine polyps, fibroids, or masses; however, endometrial biopsy and/or hysteroscopy is usually necessary for confirmation.

- Tamoxifen (common breast cancer treatment) may predispose to endometrial cancer and hyperplasia. Women on tamoxifen could be screened with US at regular intervals (controversial).
- Endometrial cancer is the most common invasive gynecological malignancy.
- Ovarian cancer has the highest mortality rate of common female cancers.
- Cervical or submucosal fibroids can impair fertility (US or MR).
- Hysterography can verify fallopian tube patency in infertility.

GU PEARLS WITH NAMES

- Krukenberg's tumor is a GI cancer (classically stomach cancer) met to the ovary.
- Stein-Leventhal disease is polycystic ovarian disease with bilateral big ovaries, infertility, and virilization.
- Fitz-Hugh—Curtis syndrome is perihepatic inflammation associated with PID.

Suggested Readings

Balthazar Emil J, ed. The radiologic clinics of North America: imaging the acute abdomen. Philadelphia: WB Saunders, 1994:32(5).

Brewer BJ, Golden GT, Hitch DC, et al. Abdominal pain. An analysis of 1000 consecutive cases in a university hospital emergency room. Am J Surg 1976;131:219.

Callen PW. Ultrasonography in obstetrics and gynecology, 3rd ed. Philadelphia: WB Saunders, 1994.

Chen MYM, Pope TL, Ott DJ, eds. Basic radiology. New York: McGraw-Hill, 1996.

Dahnert W. Radiology review manual, 2nd ed. Baltimore: Williams & Wilkins, 1993.

Dunnick NR, Sandler CM, Amis JR, Newhouse JH. Textbook of uroradiology, 2nd ed. Baltimore: Williams & Wilkins, 1997.

Ell SR. Handbook of gastrointestinal and genitourinary radiology. St Louis: Mosby Year Book, 1992.

Mirvis SE, Shanmuganathan K. Abdominal computed tomography in blunt trauma. Semin Roentgenol 1992;27(3):150–183.

O'Boyle MK. Ectopic pregnancy: an update. Applied Radiology 1996; 25(8):24–29.

Ravin CE, Cooper C, Leder RA, eds. Review of radiology, 2nd ed. Philadelphia: WB Saunders, 1994.

Rumack CM, Wilson SR, Charboneau JW. Diagnostic ultrasound. St Louis: Mosby Year Book, 1991.

Smith RC, Verga M, McCarthy S, Rosenfield AT. Diagnosis of acute flank pain: value of unenhanced helical CT. AJR Am J Roentgenol 1996; 166:97–101.

Squire LF, Novelline RA. Fundamentals of radiology, 4th ed. Cambridge: Harvard University Press, 1988.

Chapter 6

Pediatric Emergencies

Pediatric emergencies are not just mini adult emergencies. Often the **age** will tell you what to look for, and may be the most important history. For example, intussusception and croup present at 3 months to 3 years and foreign body aspiration at 6 months to 3 years. Explain tests in terms understandable to children. Radiation risk is small for most radiographic studies, but real. Ultrasound (US) or magnetic resonance (MR) may be reasonable alternatives.

CHILD ABUSE / TRAUMA

Between 1 and 2 million children are abused or neglected annually in the United States. The most commonly abused are defenseless and under 2 years of age (75% of cases). When the mechanism of reported injury doesn't match the type of injury or if the history is vague, **child abuse** must be considered. The most characteristic injuries are the growth plate "corner" or "bucket-handle" fractures, posterior rib fractures, skull fractures, subdural hematomas (Fig. 6.1), or multiple fractures in different stages of healing. Regular bone shaft fractures are four times as *common* in abuse than the more *characteristic* metaphyseal fractures (which are bone chips at the edge of the growth plate). Abdominal blunt injuries involving the duodenum and pancreas also occur. The third part of the duodenum may be bruised if it is slammed against the adjacent spine. Trauma is actually the most common cause of pediatric pancreatitis. Children are more prone to traumatic renal laceration, bronchial tear, heart contusion, or injury to the aorta or great vessels than are adults.

GROWTH-PLATE FRACTURES

The **Salter-Harris growth-plate fracture** classification is easily recalled with the acronym "**SALTR**" (Fig. 6.2).

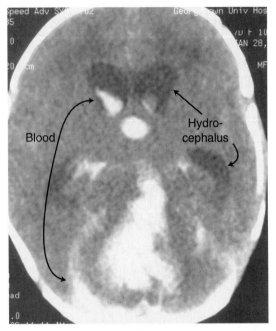

Figure 6.1. Shaken baby.

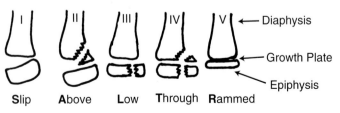

Figure 6.2. Salter-Harris growth plate fracture classification.

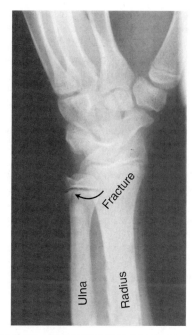

Figure 6.3.   Salter-Harris Type II fracture in ulna.

Type I = S = **S**lipped growth plate. (Usually younger than 5 years old; i.e., slipped capital femoral epiphysis.)

Type II = A = **A**bove the growth plate. (*Most common type*, approximately 70% of growth plate fractures—distal radius or tibia is a common location [Fig. 6.3].)

Type III = L = **L**ow or below the growth plate.

Type IV = T = **T**hrough the growth plate.

Type V = R = **R**ammed or crushed growth plate.

Types IV and V may need surgical reduction/fixation and may result in growth deformities, whereas types I, II, and III have excellent outcomes. (Type III may need closed reduction.)

FRACTURES

Typical **childhood fractures** include torus or **buckle fractures,** where only a subtle buckling of the cortex is apparent without a lucent fracture line. In a **greenstick fracture,** the softer young bone breaks like a freshly-cut green stick; the break only goes partway through the bone shaft, leaving the periosteum intact. Soft growing bone may also bend without breaking, causing a bowing or **plastic fracture.** A **toddler's fracture** is a spiral hairline fracture usually in the mid to distal tibia in infants 1 to 2 years old.

The presence of a posterior fat pad on the lateral elbow film signifies a **supracondylar fracture of the humerus** in a younger child, and a radial head fracture in an older child or adult (Fig. 7.5). A line drawn along the anterior humerus on the lateral elbow film should run through the middle third of the capitellum, in the absence of a supracondylar humerus fracture (Fig. 6.4). In many cases, examining the patient for point tenderness or even comparison films of the normal opposite side may resolve ambiguity. X-rays are usually not helpful with the **nursemaid's elbow,** where the radial head is dislocated by pulling on an infant's arm. Supination and flexion at the elbow successfully reduce it, rarely causing an avulsion or chip fracture.

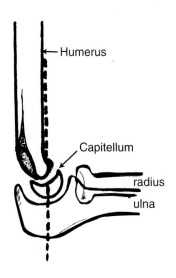

Figure 6.4. Lateral elbow (humerus): anterior humeral line normally goes through middle third of capitellum ossification center.

BONE

Bone trauma or infection (**osteomyelitis**) may be harder to diagnose in small children who can't tell you a history. Early osteomyelitis is usually seen on nuclear medicine bone scan within 2 days of injury or infection, whereas the less sensitive x-rays can take 2 weeks to show changes of infection. A septic joint may show joint space widening with a joint effusion. A septic hip effusion or congenital hip dislocation can be diagnosed with US, which can evaluate the hip joint in children under 1 year old.

- Sickle cell anemia predisposes to bone infarcts (osteonecrosis), and may present with swollen fingers and toes at almost 2 years old (hand-foot syndrome).
- Thalassemia predisposes to marrow hypertrophy and redistribution (causing a thick skull).

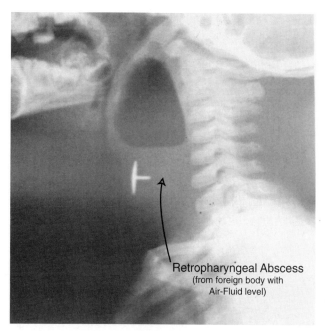

Retropharyngeal Abscess
(from foreign body with
Air-Fluid level)

Figure 6.5.

NECK AND THROAT

Widened prevertebral soft tissue on a lateral neck film may signify **retropharyngeal abscess,** and CT will clarify (Fig. 6.5); however, over-flexion may simulate this soft tissue widening. Also, beware of "pseudosubluxation," which is when normal cervical vertebral bodies may falsely appear to be slipped or offset (listhesis), due to the looser ligaments of childhood.

Croup causes subglottic soft tissue edema usually in patients 6 months to 3 years old. It is commonly from parainfluenzae virus and results in the "steeple" sign on frontal neck or chest x-ray. This describes the abnormally-pointed, gradually tapering appearance of the upper tracheal air stripe, which normally has abrupt shoulders like a bottle of wine (Fig. 6.6). Membranous croup (staphylococcal tracheitis) occurs in older, more toxic patients than the much more common viral croup.

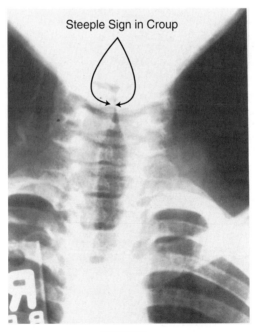

Figure 6.6.

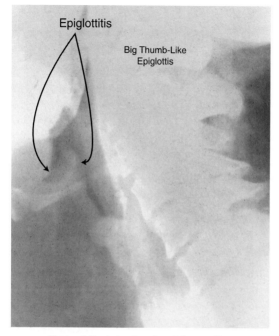

Figure 6.7.   Lateral neck soft tissues film.

Epiglottitis causes dangerous airway narrowing, as well as croup-like subglottic swelling in 25% of cases. While classically described in children 3 to 6 years old, it is more common to see it in adults now because of the *Hemophilus influenzae* vaccine. The plump thumb of the swollen epiglottis floating in the hypopharynx on the lateral neck x-ray is characteristic (Fig. 6.7). The normal epiglottis should only be pencil-thin. If there is suspicion of epiglottitis (leaning forward and drooling), invasive procedures should be avoided. A portable soft tissue neck x-ray should be considered, but evaluation and possible intubation should be performed in the operating room.

CHEST

An **aspirated foreign object** is common in children between 6 months and 3 years old (Fig. 6.8). With inspiration, there is a

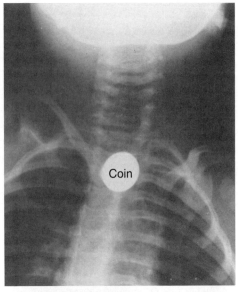

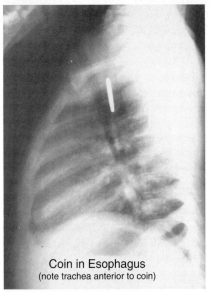

Coin in Esophagus
(note trachea anterior to coin)

Figure 6.8.   Foreign body ingestion.

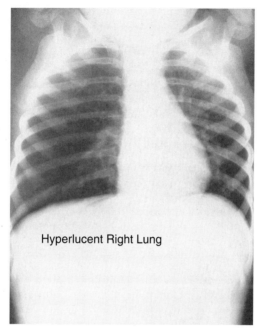

Figure 6.9. Aspirated foreign object.

shift of the mediastinum and diaphragm away from the side of the aspirated object on chest x-ray or fluoroscopy. Likewise, with expiration, there may be air-trapping with a persistent lucency on the affected side (Fig. 6.9). If necessary, decubitus films show that the obstructed lung won't shrink when dependent (down). Atelectasis and consolidation are late findings with foreign object aspiration. CT is reserved for equivocal cases. Bronchoscopy is performed if there is a high suspicion, regardless of radiographic findings, due to a high false negative rate for plain radiographs.

The child's heart occupies a relatively larger part of the thorax than in adults, but the cardiothoracic ratio is unreliable in children. The **thymus** fills in the anterior mediastinum, and may simulate a mass. The **sail and wave signs** may help identify this (see Chapter 2).

Neonatal and infant **chest** x-rays can be melted down to analysis by age, level of aeration, and nature of infiltrates. **Wet lung** (formerly transient tachypnea) of the newborn typically occurs in cesarean section babies, is completely normal by 2 to 3 days of age, and may be hyperinflated. This is contrasted with neonatal pneumonia and meconium aspiration that may also hyperinflate, but have persistent infiltrates beyond this age. **Meconium aspiration** infiltrates are coarse and patchy with occasional pneumothoraces. Immature lung (hyaline membrane disease/**surfactant deficiency syndrome**) occurs in premature infants and is underinflated with fine, hazy microatelectasis.

- **Bronchopulmonary dysplasia** may lead to pulmonary interstitial emphysema that has lucent streaks among coarse scarred lungs following chronic ventilation and barotrauma.
- Umbilical catheter tips should reside from L3 to L5 or T6 to T10, to avoid clots in large branches of the inferior vena cava and aorta (remember this with the trick "3 + 3 = 6 and 5 + 5 = 10").

Viral pneumonia or bronchiolitis in children typically has diffuse perihilar hazy interstitial infiltrates with peribronchial cuffing and hyperinflation (from reactive airways). **Round pneumonia** is common in children under 8 years old, may be mass-like, and is usually due to staphylococci or streptococci. Postinfectious **pneumatoceles** are air-filled cysts (more common in childhood pneumonias, characteristically staphylococcus). Other causes of cystic densities include posttraumatic pneumatoceles, pulmonary sequestration, bronchogenic cyst, diaphragmatic hernia, and cystic fibrosis.

Cystic fibrosis patients get recurrent upper lung interstitial pneumonias with lymphadenopathy, bronchiectasis, and mucous plugging (Fig. 6.10). Parallel train-track lines radiating from the hilum may be bronchiectasis. In cystic fibrosis patients, chest x-rays must be compared to old films to be meaningfully evaluated.

- Differential diagnosis for unilateral dark lung: endobronchial foreign body (Fig. 6.9), pneumothorax, congenital lobar emphysema, hypoplasia of the lung.
- The thymus shrinks with stress and regenerates following recovery. It may regenerate larger than baseline.
- **Down syndrome** patients are predisposed to atrioventricular canal defects, atlantoaxial instability, and duodenal atresia.

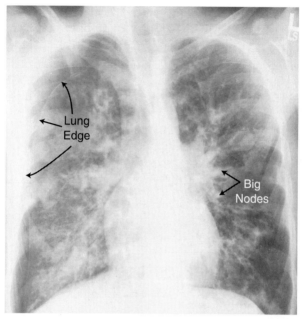

Figure 6.10. Cystic fibrosis with pneumothorax and left hilar adenopathy.

- Most common anterior mediastinal mass overall is lymphoma (Fig. 2.21). Lymphoma can present with tracheobronchial compression requiring emergent radiation therapy.
- The most common pediatric lung neoplasm is metastasis, most commonly from Wilms tumor, osteosarcoma, or Ewing sarcoma.

ABDOMINAL EMERGENCIES

Childhood **abdominal emergencies** may be divided by age at presentation. Gas in the bowel wall (pneumatosis) may signify necrotizing enterocolitis in the premature neonate with bowel distention (Fig. 6.11). The most common surgical emergency of infancy is **pyloric stenosis.** It presents between 2 and 8 months of life (males 4 to 1) with projectile vomiting (squirt-gun), a palpable pyloric mass (**olive sign**), and a distended stomach on x-ray. US is usually diagnostic. **Duodenal atresia** or stenosis presents in the

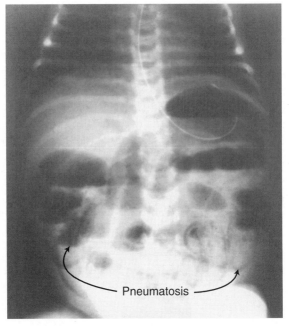

Figure 6.11. Necrotizing enterocolitis.

first days of life with bilious vomiting (yellow/green) and a "**double bubble**" on x-ray from the distended stomach and duodenal bulb. Jejunal or ileal atresias also present with high obstruction. **Malrotation with resulting small bowel volvulus** with obstructing fibrous Ladd bands is a surgical emergency and may also have a "double bubble." Diagnosis is usually by contrast study.

   Intussusception (telescoping of the bowel into itself) may present with a palpable mass; intermittent cramps; and bloody, jellylike diarrhea. It occurs from 3 months to 3 years, and is commonly ileocolic and idiopathic in children. (Adults with intussusception usually have pathologic masses that telescope as lead points.) On x-ray, there may be a small bowel obstruction or absent bowel gas in the right lower quadrant. Contrast or air enema with inside-out bowel is diagnostic and usually also therapeutic, reducing the intussusception in most cases. US is a less sensitive screening tool, showing a bull's eye or target sign on transverse images, or a "pseudokidney" sign.

Meckel diverticulum is seen in 2% of autopsies, occurs 2 feet proximal to the ileocecal valve, and has complications before the age of 2 in 2% of these patients (**rule of twos**). The main complication is gastrointestinal (GI) bleeding from ulcerating ectopic gastric mucosa, which can be identified with a nuclear medicine Meckel scan. **Hirschsprung's disease** (congenital mega-colon from denervated colon) is diagnosed with barium enema and usually presents in the first month of life. **Appendicitis** peaks in the teen years, and is rare under 7 years. US is often the first study to evaluate clinically confusing patients because it gives no radiation. CT is the most sensitive study, however (see Chapter 4).

Abdominal masses may displace adjacent structures. **Wilms** tumor of the kidney and **neuroblastoma** of the adrenal are two common early childhood tumors. Wilms lung metastasizes to the lungs. Neuroblastoma is the most common solid malignant tumor in a child younger than 4 years old, and bone metastases are common. In the neonate, hydronephrosis and multicystic dysplas-tic kidney are the number one and number two causes of abdomi-nal masses. The hydronephrosis is usually due to a congenital ureteropelvic junction obstruction. US, CT, and sometimes MR are indicated to check abdominal masses out.

- A lateral decubitus film may replace an upright film if the pa-tient is too young or unable to cooperate.
- **Leukemia, CNS tumors, and lymphoma are the 3 most common neoplasms of childhood.** (Wilms tumor and neuroblastoma are the next.)
- Rhabdomyosarcoma is the most common soft tissue sarcoma in children.

TESTICULAR TORSION

A painful scrotum is considered a **testicular torsion** until proved otherwise. The time from pain onset to surgery determines sal-vageability (best if less than 6 hours, and then the salvage rate goes down to approximately 20% at 12 to 24 hours). Acutely, nuclear medicine testicular scan is highly accurate, showing de-creased or no blood flow to the twisted testicle. Doppler US shows absent flow. (Thus, presence of Doppler flow can exclude acute torsion.) This must be differentiated from epididymitis and orchitis that have increased flow to the epididymis and testicle. After-hours, US is usually more readily available and gives a quicker answer than nuclear medicine.

BRAIN US

The fontanelle may be used as a US window in neonates to look for bleeds or hydrocephalus. This is most helpful in premature neonates because 90% of germinal matrix bleeds occur in the first week of life. US grades for **brain bleeds** in neonates predict mortality.

CONGENITAL HEART DISEASE PEARLS

• **Bicuspid aortic valve** is the most common *congenital* cardiac lesion overall, and may lead to aortic stenosis in adulthood.
• **Ventricle septal defect (VSD)** is the next most common *intracardiac* congenital heart lesion and the most common left-to-right shunt.
• **Atrial septal defect (ASD)** is the most common shunt that persists into *adulthood* (female 8:1).
• **Transposition of the great vessels** is the most common cyanotic congenital heart disease *with overcirculation* and looks like an "egg on a string" (big heart with a narrow superior mediastinum). This is also the most common *cyanotic* congenital heart disease presenting *at birth*.
• **Tetralogy of Fallot** is the most common *cyanotic* congenital heart disease of children and adults overall. This is also the most common lesion presenting with cyanosis after 1 month of life. The vascularity is decreased and the heart is **boot-shaped.**
• **Eisenmenger physiology** is reversal of flow direction in a long-standing left-to-right shunt. Chest x-ray shows big central pulmonary vessels with abrupt pruning or tapering of peripheral vessels.
• **Coarctation of the aorta** causes **rib notching** along the bottom of ribs #3 to 8 from collateral circulation (Fig. 2.16), but rarely before the teen years. The aortic arch may be small or indented with poststenotic widening (the figure 3 sign). Turner syndrome, cerebral berry aneurysms, and bicuspid aortic valve (and, thus, aortic stenosis) are associated.

Suggested Readings

ARRS Symposium. Pediatric emergencies, categorical course syllabus. American Roentgen Ray Society Annual Meeting. Washington, DC: May 1995.

Blickman JG. Pediatric radiology: the requisites. Philadelphia: Mosby Year Book, 1994.

Dahnert W. Radiology review manual, 2nd ed. Baltimore: Williams & Wilkins, 1993.

Elliott LP, ed. Cardiac imaging in infants, children, and adults. Philadelphia: JB Lippincott, 1991.

Kirks DR. Practical pediatric imaging. Boston: Little, Brown, 1984.

Ravin CE, Cooper C, Leder RA, eds. Review of radiology, 2nd ed. Philadelphia: WB Saunders, 1994.

Chapter 7

Bone/Musculoskeletal

X-RAYS

Many processes masquerade as musculoskeletal pain. Be on the lookout for common normal variants. You must be able to imagine a *realistic* potential diagnosis before ordering a study. For example, many lumbosacral spine films are done for no good treatable reason in young patients with clear-cut reasons for low back pain. In a lumbar spine x-ray series, the gonads receive the equivalent radiation dose of having a daily chest x-ray for many years. A study can be justified, however, with acute trauma in the elderly or if the symptoms have not responded to conservative management.

The practical approach to bone includes the **ABCDs:** **A**lignment (subluxed versus dislocated); **B**one mineralization (osteomalacia or osteopenia) and **B**one formation (osteophytes, periostitis); **C**ortical disruption or fracture, **C**artilage space (joint narrowing, erosions, or fusion), and **C**alcification in joint or tendons; **D**istribution (important in arthritis), and **S**oft tissue (swelling or calcification).

• *Subluxation* is a partly slipped articulation with some joint contact remaining versus *dislocation,* which has no joint contact.

BONE SCAN AND MAGNETIC RESONANCE (MR)

Some fractures won't show up on x-ray until the osteoclasts have time to resorb bone from the fracture line; thus, consider repeat films in a week if symptoms persist. **Discitis** or **osteomyelitis** is diagnosed earlier by nuclear medicine **bone scan** or MR than by x-rays. Also, bone scan can often tell superficial cellulitis from osteomyelitis. X-ray findings of infected bone include periostitis early and, later, cortical destruction. Bone scans are commonly used to screen for bone metastases in cancer patients.

MR is the primary study for early osteonecrosis; soft tissue and bone tumors; and injury to muscles, tendons, or ligaments. MR provides an exquisite window into ligamentous, tendinous, joint, and soft tissue pathology; however, even MR often needs x-ray or computed tomography (CT) correlation to better evaluate cortical bone (fractures, periostitis, or calcification patterns).

Occult processes may be hard to diagnose and include stress phenomena, nondisplaced growth plate fracture, nondisplaced hip fractures, early septic joint, discitis, or early osteomyelitis. Bone scan or follow up x-ray may be diagnostic in these settings.

PERIOSTITIS

Periostitis is a nonspecific secondary sign of bone abnormality and may indicate tumor, infection, fracture, or systemic illness. A **Codman's triangle** describes the periosteum elevated off the cortex by an aggressive process (classically osteosarcoma), and then disappearing into the tumor mass (Fig. 7.1). "Sunburst" densities radiating perpendicular to the cortex plane also represent an aggressive periostitis (Fig. 7.1). Less aggressive periostitis may show a better-defined onion-skin or layered appearance, or even diffuse cortical thickening. Less aggressive implies a slower, more likely benign, process. Likewise, well-defined borders suggest a less aggressive process, whereas ill-definition or adjacent soft tissue mass points to more aggressive entities.

Bone lesions are described as lytic (dark or lucent) or blastic (white or sclerotic) compared with normal bone. The type of bone matrix may narrow the diagnosis of bone lesions. **Osteoid** is fluffy or cloud-like calcification, whereas **chondroid** is rings, arcs, or popcorn bone formation.

FRACTURES

Be able to describe a fracture. "Transverse," "spiral," or "oblique" describes the course of the fracture line. A "comminuted" fracture has many lines, and may have unconnected "butterfly" fragments. More descriptive terms include angulation, alignment and displacement (offset), impaction (shortening) and distraction (separation), and intra-articular extension.

Fractures are treated with either immediate orthopedic evaluation (reduction and fixation), or immobilization and splinting,

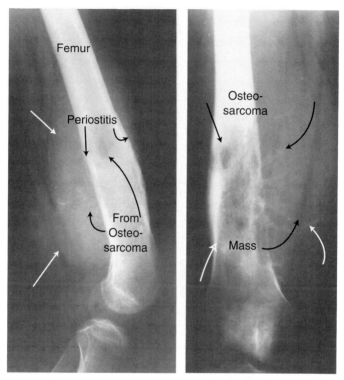

Figure 7.1. A. Lateral. B. AP, femur.

with later follow up. All injuries deserve evaluation of the neuro-vascular status distal to the fracture. Open fractures, intra-articular involvement and major alignment or position deformities may require reduction and/or casting. Complete evaluation requires at least two views of the area in question at 90 to each other (Fig. 7.2), and may include x-rays of the adjacent joints, above and below the fracture (although these may also be cleared clinically). CT may be particularly helpful in looking for fractures of the spine, acetabulum, scapula, calcaneus, and tarsal or carpal bones.

Old fractures and accessory ossification centers have bright cortical bone at all edges (well corticated). Old healed fractures

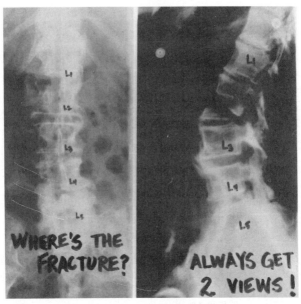

Figure 7.2. AP and lateral lumbar spine.

develop bony overgrowth across the fracture line (callus). New fractures have sharp edges without a bright cortex. Consult a normal variants text to see the many variants that simulate fracture. A small chip of bone pulled off at the insertion site of a tendon or ligament is termed an **avulsion** fracture (Fig. 7.3).

Some fractures only have indirect and nonspecific signs. A **lipohemarthrosis** in the knee is a horizontal fat-fluid level from liberated bone marrow fat and blood mixing with synovial fluid (Fig. 7.4). It can be seen with tibial plateau fractures. **Fat pads** are lucent (dark) and may be created or displaced with fracture. On a lateral elbow x-ray, the presence of a posterior fat pad signifies effusion and/or fracture, regardless of whether the fracture itself is seen. The effusion will also cause the normally seen anterior fat pad to bulge, making it look like a triangular sail (the sail sign) (Fig. 7.5). This is usually from a radial head fracture in adults.

It is more precise and less confusing to simply describe the

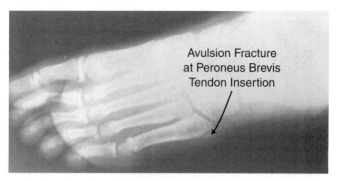

Figure 7.3.   Avulsion.

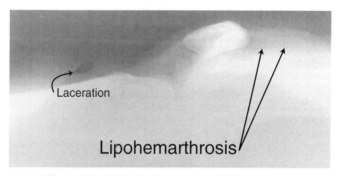

Figure 7.4.   Hidden fracture on cross-table lateral knee.

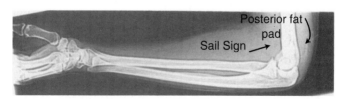

Figure 7.5.   Fat pads indicate hidden fracture.

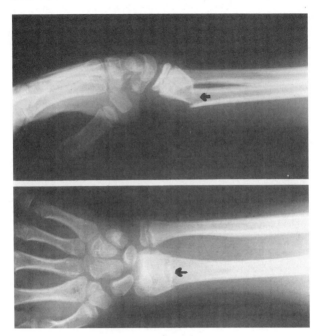

Figure 7.6. Colles fracture.

fracture and avoid using the ambiguous names. Here are a few of the more common useless names:

UPPER EXTREMITY FRACTURES WITH NAMES

- **Colles-type**—fracture of the distal radius with dorsal or neutral tilt of the distal radius joint surface (from a fall on an outstretched hand) (Fig. 7.6).
- **Hill-Sach's** deformity—fracture pit in the superior-posterior-lateral aspect of the humeral head (from repeated anterior shoulder dislocations) (Fig. 7.7).
- Bankart lesion—fracture or pit of the anteroinferior margin of the glenoid rim (also from repeated anterior shoulder dislocations).
- Nightstick fracture—solitary transverse fracture of the ulna shaft (from direct trauma).
- **Scaphoid** fracture—snuffbox tenderness on examination. Frac-

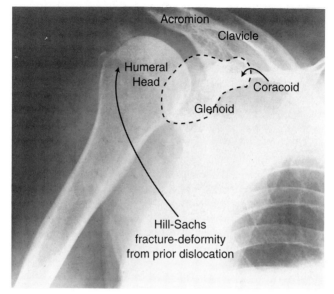

Figure 7.7.

ture may only show up on delayed x-ray. May be complicated by osteonecrosis of the proximal fragment of the scaphoid, as the blood supply may enter distally and curve back retrograde (disrupted by fracture) (Fig. 7.8). Similar in talus fractures.

- **Boxer's**—fracture of the distal fifth metacarpal (from a fist slammed down on a table, or a fist that doesn't know how to box correctly) (Fig. 7.9).
- Bennett's—simple fracture/dislocation of the base of the first metacarpal (thumb knuckle). Oblique fracture line involves the carpo-metacarpal joint with proximal dislocation of the metacarpal shaft.
- **Galeazzi's**—fracture of the radius at the junction of the mid and distal thirds with associated subluxation of the distal ulna (Fig. 7.10).
- **Monteggia**—fracture of the proximal third of the ulna with associated anterior dislocation of the radial head.

("**MUGR**" = **M**onteggia has **u**lna fracture, **G**aleazzi has **r**adius fracture.)

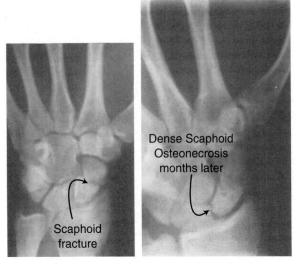

Figure 7.8.

LOWER EXTREMITY FRACTURES WITH NAMES

- **Jones (or dancer's)**—transverse fracture of the lateral base of the fifth metatarsal or an avulsion fracture at peroneus brevis insertion site (Fig. 7.3), usually from high heels, flexion, and inversion (versus accessory growth plate, which has sclerotic or better defined margins).
- **Segond's**—avulsion fracture of the ileo-tibial band insertion site at the lateral margin of the upper tibia. Associated with anterior cruciate ligament tear (MR is the next study).
- **Lisfranc**—fracture/dislocation of tarso-metatarsal joints.
- **March**—stress or fatigue fracture of metatarsal heads.

OSTEONECROSES OR AVASCULAR NECROSES (AVN) WITH NAMES

- **Legg-Calve-Perthes' disease**—idiopathic femoral head in child.
- **Osgood-Schlatter disease** (surfer's knee)—tibial tuberosity in adolescent males.
- **Femoral head** (Fig. 7.11)—scaphoid, talus, proximal tibia also predisposed.

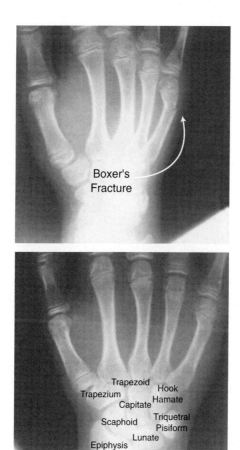

Figure 7.9.

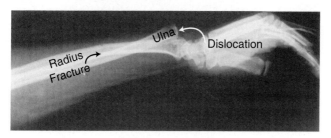

Figure 7.10.   Galeazzi.

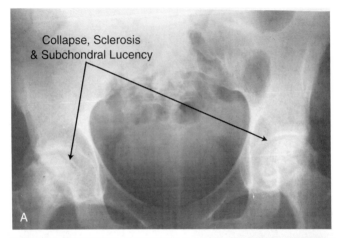

Collapse, Sclerosis
& Subchondral Lucency

A

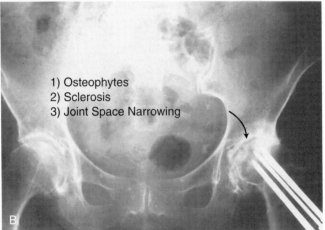

1) Osteophytes
2) Sclerosis
3) Joint Space Narrowing

B

Figure 7.11.   A. Osteonecrosis femoral heads. B. Degenerative arthritis.

TRAUMA PEARLS

- Calcaneus fracture is associated with lumbar spine compression fracture.
- Pelvic fractures often occur in pairs, as the pelvis forms a bony ring—at least two sides of the ring usually fracture, just as if you snapped a pretzel in half. Pelvic fractures are best evaluated with CT.
- In kids, elbow fractures are usually supracondylar humerus, while adults suffer radial head fractures more frequently. (Whereas the same fall on an outstretched hand will cause a Colles' fracture in the elderly.)
- **Mallet finger** (baseball finger)—flexion deformity distal phalanx due to extensor tendon avulsion.

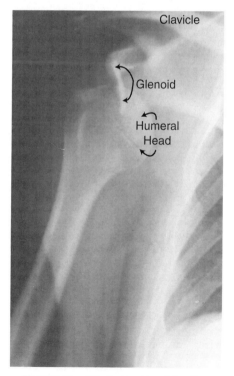

Figure 7.12.   Anterior shoulder dislocation.

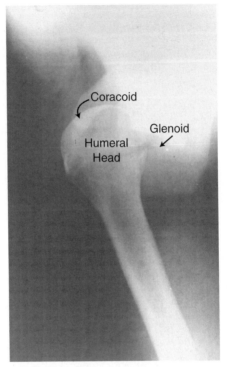

Figure 7.13. Anterior dislocation on axillary-Y view. (Humeral head slides toward coracoid.)

- **Shoulder "dislocations"** are described by the position of the humeral head in relation to the glenoid, and are almost always anterior (Fig. 7.12). The much less common posterior dislocations are usually due to seizure, electrical injury, or high-impact trauma. Scapula-Y or axillary-Y views can differentiate these from normal (Fig. 7.13).
- **Shoulder "separations"** are subluxations or dislocations across the acromioclavicular joint and don't require reduction.
- **Rotator cuff tears** involve the supraspinatus tendon in the avascular zone 1 cm from the tendon insertion (baseball pitchers and rheumatoid arthritis patients).

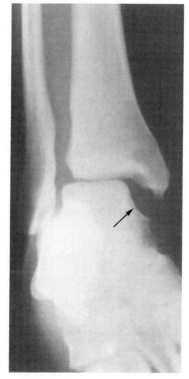

Figure 7.14. Disrupted ankle mortis.

- David Letterman sign—**scapholunate dissociation** where the space between the scaphoid and lunate is wide (normal is less than 2 mm) simulating the space between Letterman's front teeth.
- The radius, lunate, and capitate should line up on a lateral wrist film.
- A bone chip seen dorsal to the wrist bones on a lateral film represents a **triquetral fracture**
- **Terrible triad**—torn medial collateral ligament and medial meniscus (connected), plus torn anterior cruciate ligament.
- Ankle x-rays may be done for ankle sprains with point tenderness or inability to bear weight (Fig. 7.14)

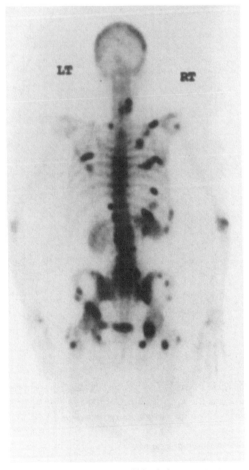

Figure 7.15. Bone scan of blastic bone metastases.

MISCELLANEOUS PEARLS

- The most common bugs to cause bone infection are staphylococcus and streptococcus.
- Nuclear medicine offers **bone densitometry,** which can quantify osteoporosis.
- Fishmouth, H-shaped, or "Lincoln-log" vertebral bodies occur in **sickle cell** anemia (bone infarcts).
- **Hyperparathyroidism** causes scalloped or bevelled edges along the radial aspect of the middle phalanges (from subperiosteal bone resorption).
- **Renal osteodystrophy = osteomalacia** (poorly mineralized chalky bones) + **hyperparathyroidism.** Can also have striped vertebrae (rugger jersey spine) with osteosclerosis.
- Joint findings in **osteoarthritis** (degenerative arthritis): osteophytes (overgrowth) (Fig. 4.6), sclerosis (white bone at joint margin), joint space narrowing (Fig. 7.11).

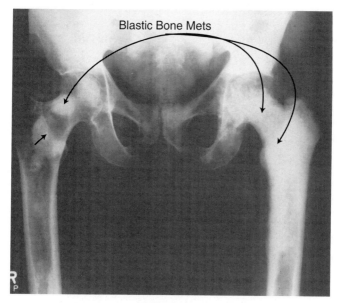

Blastic Bone Mets

Figure 7.16.

BONE TUMOR PEARLS

- The most common **primary bone tumors** are: **#1: multiple myeloma** (older patients only) and **#2: osteosarcoma** (younger patients, but bimodal), followed by chondrosarcoma and Ewing sarcoma.
- **"Pathologic"** **fractures** occur in abnormal bone across bone lesions (primary tumors, metastases, and benign cysts).
- **Metastases** like to go to the more central skeleton (spine pedicles).
- 4 out of 5 **bone malignancies** can be correctly diagnosed by **age** alone:

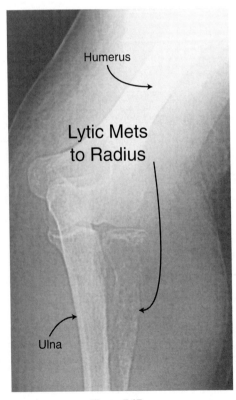

Figure 7.17.

younger than 1 month = neuroblastoma
1 to 15 yrs = Ewing, Langerhan cell histiocytosis (eosinophilic granuloma)
10 to 30 yrs = osteosarcoma
30 to 40 yrs = lymphoma, fibrosarcoma
older than 40 yrs = metastases, multiple myeloma, chondrosarcoma

DIFFERENTIAL DIAGNOSIS (DDX) MNEMONICS

- **Blastic** (dense) **metastases:** "**5 B**ees **L**ick **P**ollen"—**B**reast (can be mixed lytic and blastic), **B**owel, **B**ronchus (carcinoid), **B**ladder (can be either lytic or blastic), **B**rain (medulloblastoma), **L**ymphoma, and **P**rostate (Figs. 7.15 and 7.16).
- **Lytic** (lucent) **metastases:** "**BLT** sandwich with a **K**osher **Pic**kle"—**B**reast, **L**ung, **T**hyroid, **K**idney, and **P**rostate (more commonly blastic) (Fig. 7.17).
- **Hypervascular metastases** (classically thyroid, kidney, and breast) may be lytic and expansile ("blow-out" lesion).
- **Osteonecrosis** or "**ASEPTIC**" necrosis—no blood supply, so bone dies; due to: **A**lcohol, **S**ickle cell, **S**teroids, **E**mbolism, **P**ancreatitis, **T**rauma, **I**diopathic, and **C**ollagen vascular disease (lupus) (Fig. 7.11).

Suggested Readings

Chen MYM, Pope TL, Ott DJ, eds. Basic radiology. New York: McGraw-Hill, 1996.

Daffner RH. Clinical radiology: the essentials. Baltimore: Williams & Wilkins, 1993.

Dahnert W. Radiology review manual, 2nd ed. Baltimore: Williams & Wilkins, 1993.

Helms CA. Fundamentals of skeletal radiology. Philadelphia: WB Saunders, 1989.

Meschan I, Ott DJ. Introduction to diagnostic imaging. Philadelphia: WB Saunders, 1984.

Ravin CE, Cooper C, Leder RA, eds. Review of radiology, 2nd ed. Philadelphia: WB Saunders, 1994.

Resnick D, Niwayama G. Diagnosis of bone and joint disorders, 2nd ed. Philadelphia: WB Saunders, 1988.

Squire LF, Novelline RA. Fundamentals of radiology, 4th ed. Cambridge: Harvard University Press, 1988.

Weist P, Roth P. Fundamentals of emergency radiology. Philadelphia: WB Saunders, 1996.

Chapter 8

Vascular/Interventional Radiology and Intravenous Contrast

Part I   Radiology

Interventional radiology/angiography can be both diagnostic and therapeutic. Procedures are done with **fluoroscopy, ultrasound, or computed tomography (CT) guidance.** Catheters or needles are inserted percutaneously and manipulated percutaneously, with guidance by real-time imaging "road maps." Imaging shows the exact location of the tip of the catheter in relation to the highway map of vessel branches, kidneys, biliary tubes, or organs. **Uses** include diagnosis and treatment of mass, abscess, clot, aneurysm, arteriovenous malformation (AVM), arteriovenous fistula (AVF), arterial narrowing, gastrointestinal (GI) bleeding, pulmonary embolus (PE), hydronephrosis, and biliary obstruction, as well as biopsy of masses or organs. Interventional radiology is often easier than the surgical alternatives in terms of risks, complications, pain, recovery times, and cost.

Intravenous (IV) sedation and pain medications (midazolam and fentanyl) take the place of general anesthesia, and many procedures are now done as outpatients. Even central venous catheters or peripherally-inserted central (PIC lines) venous catheters can be placed in interventional radiology instead of surgery. The same preoperative labs and orders should be done prior to an interventional procedure as for surgery (consent; blood urea nitrogen (BUN)/creatinine; PT/PTT; complete blood cell count (CBC); eating nothing after midnight; as well as antibiotics on-call to some kidney and biliary procedures).

ANGIOPLASTY

Blowing up a balloon on a catheter to open a blocked or narrowed vessel is called **balloon angioplasty.** This is sometimes an alternative to bypass surgery. Angioplasty is more successful in bigger arteries, and has a high success rate in the iliac arteries. Angioplasty has also been used to treat cerebral vasospasm, a common sequela of subarachnoid hemorrhage. Other applications include unclogging dialysis arteriovenous grafts and dilating esophageal or biliary narrowing. Expandable metallic or plastic tubes called **stents** may keep a vessel or other natural conduit open if repeated balloon attempts fail.

THROMBOLYSIS WITH UROKINASE

In **thrombolysis** for an ischemic limb, **urokinase** is infused through an arterial catheter placed at the clot site to try to dissolve it. This only works with fairly fresh clots. Urokinase can also be used to dissolve a clotted central venous catheter or dialysis graft. Some dialysis catheters are kept open with 5000 units of urokinase. Occasionally, urokinase may be instilled into a chest tube or drainage catheter to break up a clotted hematoma, empyema, abscess, or hemothorax.

EMBOLIZATION

Selective purposeful clotting (**embolization**) of vessels can control bleeding in GI bleeds, vascular tumors, or trauma. It is the preferred treatment for uncontrolled traumatic bleeding in the pelvis and retroperitoneum. CT is an ideal screening exam for bleeding from penetrating or blunt trauma, but angiography may be required to fully evaluate for vessel injury.

Pulmonary AVMs and recurrent hemoptysis from cystic fibrosis are other indications for embolization. Embolization may be indicated for central nervous system (CNS) AVMs (Fig. 8.1), some aneurysms, and preoperatively for meningiomas or other neoplasms.

PE, DEEP VENOUS THROMBOSIS, AND INFERIOR VENA CAVA FILTERS

Pulmonary angiography is the gold standard for PE diagnosis. A catheter is placed into the pulmonary arteries from the common

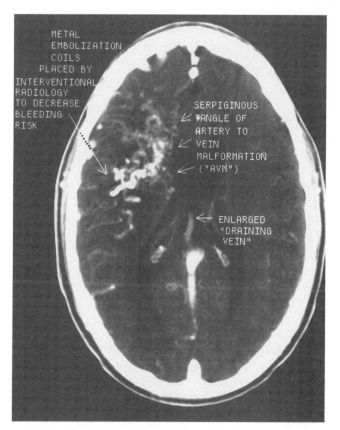

METAL
EMBOLIZATION
COILS
PLACED BY
INTERVENTIONAL
RADIOLOGY
TO DECREASE
BLEEDING
RISK

SERPIGINOUS
TANGLE OF
ARTERY TO
VEIN
MALFORMATION
("AVM")

ENLARGED
"DRAINING
VEIN"

Figure 8.1.

femoral vein, and x-rays are taken while contrast is injected into the pulmonary arterial tree. The nuclear medicine V/Q may help localize suspicious areas, and is usually done before angiography (see Fig. 2.27). Helical CT is also becoming a proven way to find central PE (Fig. 8.2).

Ultrasound of the lower extremities can look for deep venous thrombosis (**DVT**) as a source for PE. DVT is easier to diagnose and is more symptomatic above the calf. Most patients will not be treated with anticoagulants (heparin or warfarin) for superficial

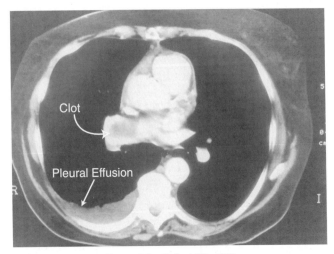

Figure 8.2. Helical CT of PE.

thrombophlebitis or if the clots are below the knee and the popliteal veins. Venography can diagnose clot if the ultrasound is equivocal.

An **inferior vena cava (IVC) filter** (Fig. 4.1) is a metallic wire mesh inserted through a femoral vein catheter into the IVC. The filter functions as a vascular umbrella, catching any big DVTs that might chip off and embolize to the lungs. IVC filters may be needed if the patient has a DVT and cannot be anticoagulated for whatever reason, if there are recurrent PEs despite anticoagulation, or if a hypercoagulable state exists (such as protein S, protein C, or antithrombin III deficiencies, anticardiolipin antibodies, or paraneoplastic syndromes). If filters break, fragments rarely can dislodge and embolize. Most filters are permanent, but removable filters are available. Long-term complications of permanent IVC filters have not been completely studied.

GI BLEEDS

Endoscopy can almost always identify an upper GI source of bleeding if done within 6 hours of admission. In life-threatening **GI bleeds,** vasopressin infusion and selective embolization may be indicated if medical or surgical therapy is unsuccessful or inappropriate. Bleeding must be brisk (greater than 0.5 to 1 cc/min) and active enough to see a blush on **angiography.** A **nuclear medicine GI bleed scan** may direct the angiographic study to the correct vessel, in the same way a V/Q scan may direct the pulmonary angiogram to the correct lung vessel. GI bleeding scans need only greater than 0.1 cc/min, and are thus 10 times more sensitive than angiography for slow active bleeding. For intermittent bleeders, the patient is injected with a radiopharmaceutical that stays in the bloodstream for approximately 24 hours, so nuclear medicine imaging may quickly be done when the bleeding recurs (see Chapter 4).

GROIN BLEEDS

Ultrasound may diagnose pseudoaneurysm or AVF following femoral artery catheterization, and CT may diagnose retroperitoneal extension of hematoma. A large amount of blood loss can occur into the retroperitoneum without being apparent on physical examination. Groin pseudoaneurysms may be treated with compression by the ultrasound transducer over the thin pseudoaneurysm origin (or "neck").

AORTIC INJURY

Aortic rupture, dissection, aneurysm, and pseudoaneurysm are all different entities. **Aortic rupture or laceration** occurs in the setting of deceleration injury. **Plain film signs** of aortic injury include an apical pleural cap of blood; tracheal or nasogastric tube deviation; big aortic knob (larger than 3 cm); hemothorax; and superior mediastinal widening (larger than 8 cm) (see Fig. 2.17). A normal CT scan with normal dark mediastinal fat around the aorta practically excludes significant **aortic injury.** If the CT is equivocal or positive, **aortography** may be needed for definitive diagnosis. Aortic rupture that makes it to the emergency room most commonly occurs at the ligamentum arteriosum.

AORTIC DISSECTION

Aortic **dissection** is blood tearing through the intima, usually traveling in the outer third of the media, and separating the arterial wall layers. Stanford type A (DeBakey type I or II) involves the ascending aorta with or without the descending aorta and is a surgical emergency. Stanford type B (DeBakey type III) involves

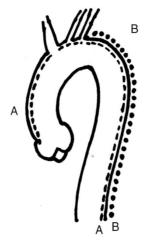

Figure 8.3. Aortic dissection types: **A** = **A**scending / **A**ll ← surgery, **B** = **B**ottom only ← medicine.

the descending aorta only and can be managed medically with beta-blockers and control of hypertension (Fig. 8.3). Remember: "**type A is for a**scending and **B is for b**ottom or descending." CT of dissection may show an "intimal flap," which will be a line through the lumen of the aorta on axial images (Fig. 8.4).

AORTIC ANEURYSM

An **aneurysm** is focal vessel dilatation (expansion of all layers of the artery wall, which remains intact). A **pseudoaneurysm** is bigger than the normal vessel, with actual disruption of one or more layers of arterial wall. If this disruption involves all three wall layers (intima, media, and adventitia), it is a contained hematoma. These are very different from dissections, but all may enlarge the aorta or mediastinum on chest x-ray.

Decisions on repair of aneurysms often depend on age and clinical presentation. Small atherosclerotic thoracic aortic aneurysms are unlikely to rupture and might not be repaired, but aneurysms due to Marfan's syndrome, syphilis, or infection should be repaired even if small and asymptomatic. Patients with a remote history of trauma may have pseudoaneurysms, or contained hematomas. If there is fracture of the first or second rib,

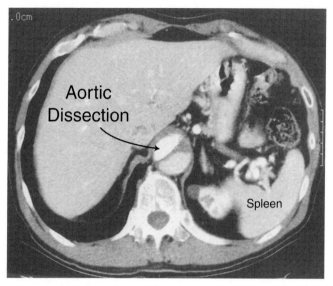

Figure 8.4. Dissection.

sufficient force has been exerted to consider arteriogram to look for aortic or subclavian tear.

An **abdominal aortic aneurysm** (AAA) is defined as larger than 3 cm in diameter. If larger than 4 cm in diameter or if it is growing faster than 5 mm per year, elective surgical repair should be considered. Aortogram, MR with MR angiography (MRA), or CT with three-dimensional reconstruction should be performed preoperatively to evaluate whether the renal, mesenteric, and iliac arteries are involved (Fig. 8.5). AAAs smaller than 4 cm may be monitored regularly for enlargement with ultrasound. Note that aortograms will only show flowing blood lumen, whereas CT and MR also show clotted blood, plaque, artery wall, thrombus, and lumen (and, thus, give a more accurate size).

RENAL ARTERY STENOSIS

Renovascular hypertension from renal artery stenosis and high renin represents less than 5% of patients with hypertension. Balloon angioplasty or stenting of this narrowing may be curative.

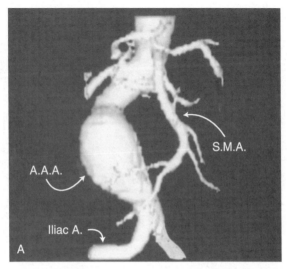

Figure 8.5. A. Three-dimensional CT arteriogram.

MRA (Fig. 8.6) may be the best screen for **renal artery stenosis;** however, CT angiography (CTA) or nuclear medicine **ACE inhibitor renal scan** are also useful screening tests. Patients typically have hypertension, renal insufficiency, and possibly a small kidney. Renal artery stenosis is usually from atherosclerosis or fibromuscular dysplasia (FMD). A kidney juxtaglomerular apparatus that is not getting enough blood makes more renin (to vasoconstrict the efferent arteriole and maintain glomerular filtration rate [GFR]). In these patients with renal artery stenosis,the efferent arteriole is dilated with angiotensin–converting enzyme (ACE)-inhibitors, and GFR is reduced, resulting in delayed uptake and cortical retention of radiopharmaceutical. This also explains why ACE-inhibitors increase the creatinine in some patients.

NEPHROSTOMY AND BILIARY DRAINAGE

Percutaneous tubes, stents, or catheters can be inserted into the biliary system or kidneys to decompress clogged plumbing from stone or tumor. A **nephrostomy** tube runs from the skin to the kidney collecting system and sometimes even continues through the

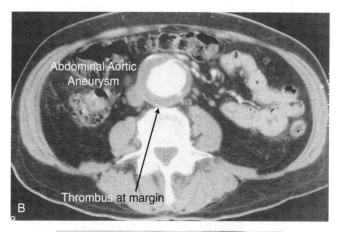

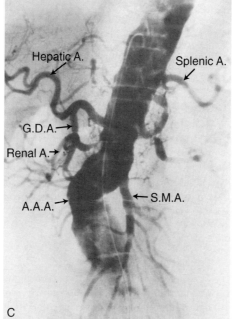

Figure 8.5. *(continued)* B. Axial CT. C. Arteriogram of same AAA.

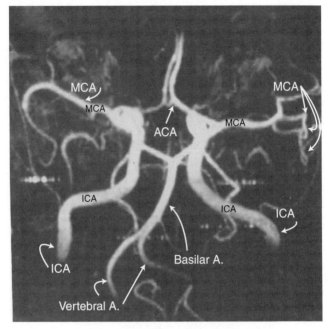

Figure 8.6.   MRA—circle of Willis.

ureter into the bladder. This allows urine to drain from the kidney to an external bag on the skin, or to the bladder if the tube is capped off. Likewise, a **biliary drain or stent** allows bile to be drained externally or to pass into the duodenum. These tubes are often placed for neoplastic blockage of the ureter or bile duct, for strictures, or for leaks (bilomas and urinomas). Surgical and laparoscopic injuries may be managed with these tubes.

OTHER PROCEDURES

Retrieval of **intravascular foreign objects** (like broken-off pieces of catheters) is done with a lasso catheter. **Transjugular intrahepatic portosystemic shunts (TIPS)** can decompress portal hypertension with placement of metallic tubes or stents that bypass the liver by rerouting blood from portal to hepatic veins. Percutaneous CT or ultrasound-guided procedures include: **abscess drain-**

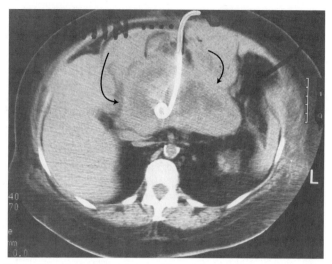

Figure 8.7.   Percutaneous drain in phlegmon/abscess from chronic pancreatitis.

age (Fig. 8.7), **mass or node biopsy, paracentesis, thoracentesis, and gastrostomy** (also done by gastroenterologists and surgeons).

NUCLEAR MEDICINE CARDIAC STRESS TEST

Nuclear medicine thallium and sestamibi heart studies evaluate coronary artery perfusion and **ischemia.** They are more sensitive and specific for coronary artery disease than an exercise stress test. Images are taken both during exercise and at rest. A "reversible defect" corresponds to an area of ischemic myocardium at risk for infarction. The "defect" refers to the lack of thallium uptake during exercise. The subsequent uptake of thallium (or "reversal") in this same area during rest differentiates ischemic myocardium from dead myocardium or scar. Thallium may also identify "hibernating myocardium" (near-dead tissue) that may work again if reperfused with bypass or angioplasty. Some departments routinely report ventriculograms with the ischemia studies to look at contractility and ejection fractions.

ANGIO / INTERVENTIONAL / CARDIOVASCULAR PEARLS

• Angiography is low-risk. 1 in 1000 have complications requiring hospitalization, 4 in 100 have minor or major complications, and the mortality rate is less than 1 in 20,000.

• Ideally, the **PT** should be less than 15 and the **PTT** should be less than 1.2 times normal.

• **Platelet** count should be greater than 60,000 to 75,000 for major interventional procedures.

• In the heparinized patient, stop heparin 2 to 4 hours prior to the procedure, if possible. **Heparin** can be reversed with protamine sulfate in an emergency. (Protamine, however, given to a diabetic on NPH insulin can cause fatal anaphylaxis.)

• **Coumadin** should be stopped days prior to elective arterial puncture (controversial). If necessary, Coumadin can be reversed with fresh-frozen plasma (this takes hours) or vitamin K (this takes days).

• Isolated gastric varices without esophageal varices suggest splenic vein clot.

• Transvenous pacing is required prior to pulmonary angiography in patients with **left bundle branch block** due to the risk of complete heart block.

• Coronary calcium seen on chest x-ray or fluoroscopy in males under age 50 is sensitive for coronary artery disease.

• Major lower GI bleeds are commonly due to diverticular disease or angiodysplasia in the right colon. However, **the most common cause of blood per rectum is an upper GI bleed!**

Part II   Contrast

IONIC VERSUS NONIONIC CONTRAST

The administration of iodinated IV contrast for CT, IVP, or angiography is not without risk. The two types of contrast classes are slightly different. **Ionic contrast** is high in osmolarity and causes more reactions, more diuresis, more renal problems, and more fatalities than its much more expensive, **nonionic,** low-osmolar counterparts. **Indications** for use of **nonionic contrast** include: allergy to iodine, shellfish, or contrast, a history of anaphylaxis, asthma, any serious drug or environmental allergy, over age 65, coronary artery disease, congestive heart failure, beta-blocker use, the potential to get more than 150 cc of contrast or an angiogram within 24 to 48 hours, or patient request. Some protocols also include diabetes, renal insufficiency, arrhythmia, multiple myeloma, sickle cell anemia, myelodysplastic syndrome, and myasthenia gravis.

CONTRAST ALLERGY AND STEROID PREMEDICATION

If there is a history of **renal insufficiency or anaphylactic (anaphylactoid) reaction to contrast,** consider doing an alternative study (unenhanced CT, ultrasound, or MR with gadolinium). Make sure the benefits of contrast use outweigh the risks (as in emergency suspicion for aortic dissection). Some CT studies do not require IV contrast to be diagnostic (abscess, kidney stone). If the study must be done emergently, then use nonionic contrast. If it is not an emergency, premedicate with steroids for contrast allergy. Contrast reactions can be fatal.

Indications for premedication with steroids include any allergy to contrast, history of major allergy or anaphylaxis, asthma, and even the question of allergy to iodine or shellfish. In these cases, the use of nonionic contrast further decreases the risk of reaction. One protocol for prophylactic steroids gives 32 mg of methylprednisolone orally 12 hours and again at 2 hours before IV contrast administration, with or without diphenhydramine and cimetidine.

CONTRAST REACTIONS AND COMPLICATIONS

The vast majority of acute reactions occur within 20 minutes of injection. If the patient has a contrast reaction, the vital signs will help differentiate a **vasovagal reaction** (fainting) from a true anaphylactic-like allergic reaction. The blood pressure goes down with both, but the **heart rate** usually goes up with anaphylaxis and down with vasovagal reaction. Vasovagal reactions are treated with Trendelenburg positioning, IV fluids, and possibly IV atropine (1 to 2 mg).

Major reactions include bronchospasm, facial or laryngeal edema, pulmonary edema, shock, and cardiac or respiratory arrest. IV fluid, oxygen, albuterol inhaler, Benadryl, epinephrine, defibrillator, and a crash cart should be close by. **Epinephrine** (0.1 to 0.3 mL of 1:1000 concentration, subcutaneous) is needed for severe respiratory compromise (bronchospasm, airway, or laryngeal edema). If hypotension is also present, epinephrine must be given **IV** (1 to 3 mL of 1 : 10, 000 concentration) or via endotracheal tube (dilute 3 times dose in 10 cc saline).

Contrast can elicit pheochromocytoma hypertensive crisis, myasthenic crisis, sickle cell crisis, or thyroid storm. **Renal damage** may result from contrast use with hypertension, diuretic use, the elderly, CHF, diabetes, proteinuria, multiple myeloma, or **creatinine** greater than 1.5. **Patients should be well-hydrated prior to receiving contrast,** especially in multiple myeloma, diabetes, and renal insufficiency. In the normal patient population, the risk for contrast-induced nephropathy is less than 2%, and is almost always brief, mild, and reversible. Reversible **acute tubular necrosis** (ATN) is the most common contrast-induced kidney problem, peaks at 3 to 5 days and improves by 7 to 10 days. Diabetes is a risk factor for renal problems from contrast if pre-existing renal insufficiency exists or if the patient is dehydrated.

CONTRAST PEARLS

- Fatality with ionic contrast is 1 in 40,000 versus 1 in 150,000 to 250,000 with nonionic. Serious reactions occur in 0.22% with ionic patients and in 0.04% with nonionic patients.
- If the patient is already in irreversible end-stage renal failure, then contrast may be given if the patient will be dialyzed.
- Checking creatinine is a safe precaution in elective studies (don't wait in emergent studies). The creatinine may transiently elevate following contrast.
- There is a dose limit for iodine contrast, therefore do only important contrast studies in the emergency setting (maximum approximately 4 mL/kg ionic, approximately 6 mL/kg nonionic).

Suggested Readings

Dahnert W. Radiology review manual, 2nd ed. Baltimore: Williams & Wilkins, 1993.

Elliott LP, ed. Cardiac imaging in infants, children, and adults. Philadelphia: JB Lippincott, 1991.

Kadir S. Current practice of interventional radiology. Philadelphia: BC Decker, 1991.

Kadir S. Diagnostic angiography. Philadelphia: WB Saunders, 1986.

Kandarpa K, ed. Handbook of cardiovascular and interventional radiologic procedures. Boston: Little, Brown, 1989.

Katayama H, Yamaguchi K, Kozuka T, et al. Full-scale investigation into adverse reaction in Japan. Risk factor analysis. The Japanese Committee on the safety of contrast media. Invest Radiol 1991;26S:S33–S36, S40–S41.

Lasser EC, Berry CC, Talner LB, et al. Pretreatment with corticosteroids to alleviate reactions to intravenous contrast material. N Engl J Med 1987;317:845–849.

Osborne A. Introduction to cerebral angiography. Hagerstown, MD: Harper & Row, 1980.

Ravin CE, Cooper C, Leder RA, eds. Review of radiology, 2nd ed. Philadelphia: WB Saunders, 1994.

Taveras J, Ferrucci JT, eds. Radiology: diagnosis, imaging, intervention. Philadelphia: JB Lippincott, 1988.

Wojtowycz M. Handbook of interventional radiology and angiography, 2nd ed. St Louis: Mosby Year Book, 1995.

Chapter 9

HIV Imaging

Patients with human immunodeficiency virus (HIV) are susceptible to a wide variety of opportunistic infections and characteristic
neoplasms, as well as non HIV-associated pathology. The **CD4
count** correlates well with this type of disease. For example, *Pneumocystis carinii* pneumonia (PCP) tends to occur with a CD4 less
than 200, whereas lymphoma, *Mycobacterium avium-intracellulare*
(MAI), cytomegalovirus (CMV), pneumatosis, and intussusception are usually late-stage phenomena (CD4 less than 50).

Emergency evaluation often includes chest x-ray and head
computed tomography (CT). Chest CT and head magnetic resonance (MR) may be helpful in complicated or confusing cases.
CT is the modality of choice to evaluate acquired immunodeficiency syndrome (AIDS)-related abdominal pathology.

Remember that patients with HIV also get the same diseases
and infections as immunocompetent patients. Community-
acquired **pneumococcal pneumonia** remains the most likely lobar
pneumonia in HIV-positive patients with near-normal CD4
counts.

CHEST

The patient with HIV and interstitial disease may have PCP, scarring from old infection, heart failure from HIV-associated cardiomyopathy, regular viral or bacterial pneumonia, or fluid overload.
Old films become important here. Diffuse infiltrate may be
congestive heart failure (CHF) or PCP, whereas focal segmental
infiltrate suggests bacterial pneumonia. Another clue is a large
heart, which points toward HIV-associated cardiomyopathy.

Most patients with HIV will get **PCP** at some time. PCP has a
variety of faces. It often begins as a fine interstitial, dusty (ground-
glass), or reticulogranular opacity. It may turn patchy or alveolar
but is usually diffuse and symmetric. In patients on inhaled penta-

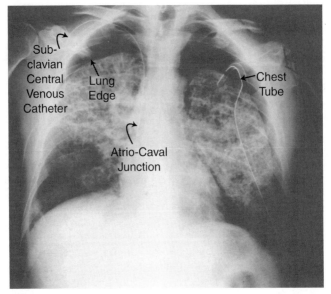

Figure 9.1. Bilateral pneumothoraces. Interstitial disease from *Pneumocystis carinii* pneumonia (PCP).

midine prophylaxis, PCP tends to infect the upper lobes. Pneumothoraces or pneumatoceles may occur in PCP (Fig. 9.1.); however, **lymphadenopathy and effusions are very uncommon in PCP, and should suggest another diagnosis.** Nuclear medicine **gallium scan** looks for tumor or inflammation, and may be positive with early PCP before the chest x-ray is abnormal.

Pleural effusions are much more typical of **Kaposi's sarcoma,** which also has nodules that are ill-defined from bleeding. Kaposi's is the only common chest problem in AIDS that is not bright on nuclear medicine gallium scan. Intervenous drug abuse increases the risk for septic emboli, recurrent staph infections, and soft tissue abscess or cellulitis. Septic emboli may form cavities in the lung (like tuberculosis [TB]).

Nodular or cavitary densities suggest Kaposi's, septic emboli, lymphoma, fungal infection (cryptococcus, aspergillosis), or TB (Fig. 9.2). The solitary pulmonary nodule is most likely lymphoma. Interstitial pneumonia in kids with HIV is usually lympho-

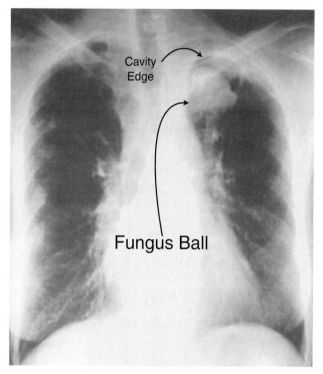

Cavity
Edge

Fungus Ball

Figure 9.2. Aspergillus in TB cavity.

cytic interstitial pneumonitis (LIP). **TB** in early AIDS resembles secondary or reactivation TB, as it is seen in the general population (infiltrates, cavitation, bronchogenic spread). TB in late AIDS is aggressive and diffuse and may resemble regular primary TB. Adenopathy and disseminated, extrapulmonary infection are more common with HIV-associated TB than with regular TB.

ABDOMEN

The gastrointestinal (GI) tract is particularly susceptible to HIV-associated illness. The **esophagus** may be infected with candida (oral thrush), herpes, CMV, or HIV itself, which can cause giant ulcers. AIDS-related **cholangitis** due to CMV or *Cryptosporidium* simulates sclerosing cholangitis, causing inflammation of the biliary tract without stones, and may cause ampullary narrowing. AIDS-associated B-cell lymphomas of the abdomen or brain are more aggressive and more likely to be disseminated or multifocal than lymphoma in immunocompetent patients. Kaposi's, TB, atypical mycobacterium (MAI), and lymphoma can all cause abdominal **lymphadenopathy,** but TB and MAI nodes are darker on CT scan. TB likes to infect the ileocecal region. One-half of homosexual patients with HIV develop Kaposi's of the GI tract.

Enteritis causing proximal small bowel dilatation and thick walls may be from *Cryptosporidium, Isospora belli, Giardia,* or MAI. *Cryptosporidium* and CMV can involve the stomach. **CMV** causes colitis and/or ileitis. Patients with HIV are also predisposed to pneumatosis and intussusception. The **pneumatosis** may be associated with free intraperitoneal air and may not signify a surgical emergency in most cases. The **intussusception** may be secondary to lymphoma, MAI, or mesenteric adenopathy acting as "lead points" for the telescoped bowel.

CT is an effective way to evaluate the symptomatic abdomen or fever of unknown origin in AIDS to search for occult lung or abdominal infection or abscess. HIV-related nephropathy causes brightly echogenic kidneys on ultrasound, and is usually a late-stage phenomenon.

BRAIN

Ring-like brightness following intravenous (IV) contrast (enhancement) in the basal ganglia is typical of **toxoplasmosis** as well as primary **lymphoma** (Fig. 9.3). Positron emission tomography (PET) scan or nuclear medicine thallium scan may differentiate the two by showing more metabolism in lymphoma. Lymphoma is more typically periventricular and solitary and toxoplasmosis is more characteristically in the basal ganglia and multifocal, but these are only slight trends with a lot of overlap. It is usually impossible to tell these two apart, and most patients receive a trial of toxoplasmosis treatment anyway.

HIV encephalopathy is the earliest and most common central nervous system (CNS) finding, with scarring and atrophy in the white matter that is symmetric and doesn't enhance. **Cryptococcal meningitis** classically causes multiple small nodules in the periventricular and inferior basal ganglia region that don't enhance.

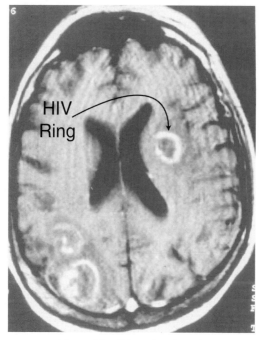

Figure 9.3.   Toxo or lymphoma?

Cytomegalovirus causes a pencil-thin high signal (bright) all around the ventricles. Progressive multifocal leukoencephalopathy (PML), from the papovavirus, is a nonenhancing, asymmetric, peripheral white matter demyelinating disease that has a high 6-month mortality. Also, remember septic emboli and TB in the CNS. Consider both typical and atypical HIV-related pathologies when imaging the patient with HIV.

Suggested Readings

Goodman PC. The chest radiograph in AIDS: infections and neoplasms. In: Thrall JH, ed. Current practice of radiology. Philadelphia: BC Decker, 1993.

Kuhlman JE. CT evaluation of the chest in AIDS. In: Thrall JH, ed. Current practice of radiology. Philadelphia: BC Decker, 1993.

Kuhlman JE, Fishman EK. Acute abdomen in AIDS: CT diagnosis and triage. Radiographics 1990;10:621–634.

Lee VW, Panageas E, Turnbull BA. Radionuclide evaluation of the chest in AIDS. In: Thrall JH, ed. Current practice of radiology. Philadelphia: BC Decker, 1993.

Sider L, Gabriel H, Curry DR, Pham MS. Pattern recognition of the pulmonary manifestations of AIDS on CT scans. Radiographics 1993; 13:771–784.

Wood BJ, Zeman RK, Cooper C, et al. Unusual abdominal manifestations of AIDS in adults. Scientific paper at American Roentgen Ray Society Annual Meeting, Washington, DC, 1995.

Chapter 10

Cancer Imaging

PATTERNS OF METASTATIC DISEASE

Patterns of metastatic disease can guide you to the diagnosis of neoplasms and their complications. **Lymphoma** is called the great imitator because of its many faces. It can be infiltrating, nodular, or mass-like in just about any organ system. If it presents as a nodal mass, it generally is soft and is deformed by adjacent structures. Most other masses tend to exert more "mass effect" by displacing or indenting bowel, lung, vessels, or whatever lives next door (Fig. 10.1). **Hodgkin** lymphoma usually presents above the diaphragm in the thorax and spreads contiguously, whereas **non-Hodgkin** lymphoma usually presents in the abdomen below the diaphragm and, more commonly, spreads to remote locations.

    Breast and **lung cancer** are the most common causes of death from malignancy. **Colorectal cancer** is the next most common, and tends to recur at the site of resection or of the original tumor (Fig. 10.2). Colon cancer tends to metastasize to the liver first, but may also go to the lung (cannonball metastases). Rectal cancer (Fig. 10.3) metastases may skip the liver (bypassing the portal venous drainage). **Hepatomas** commonly occur in cirrhotic livers and invade the portal vein (Fig. 10.4). **Renal cell carcinoma** typically grows into the renal vein and inferior vena cava (IVC) (Fig. 10.5). **Testicular** and ovarian lymphatic metastases may skip the pelvis and appear in the para-aortic or renal hilar regions. **Ovarian carcinoma** is typically cystic, can get as big as the abdomen, and metastasizes to the peritoneum, causing omental caking. **Gastrointestinal (GI) cancers** can also metastasize to the peritoneum. **Prostate and bladder cancer** spread to the regional pelvic nodes first. **Melanoma** and **breast cancer** can metastasize just about anywhere (Fig. 10.6).

PREMALIGNANT OR HIGH-RISK STATES

"Premalignant" or **high-risk states** include: ulcerative colitis (greater than Crohn disease), calcified or porcelain gallbladder,

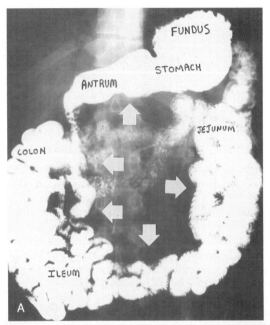

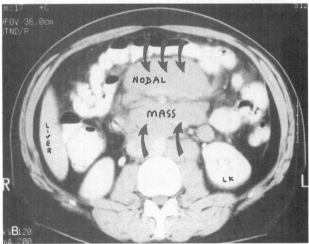

Figure 10.1. A. Mass effect from lymphoma on small bowel follow-through. B. Same lymphoma nodes on CT.

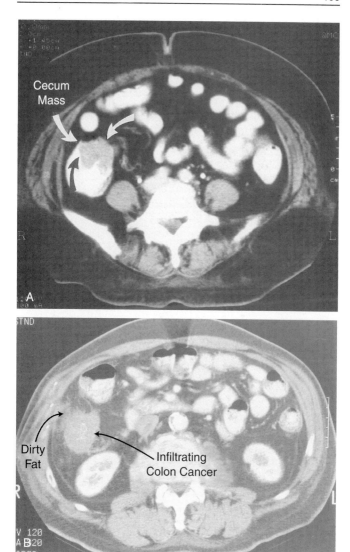

Figure 10.2. Colon cancer.

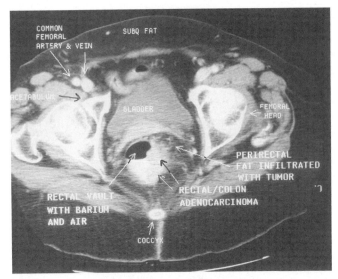

Figure 10.3. Rectal cancer.

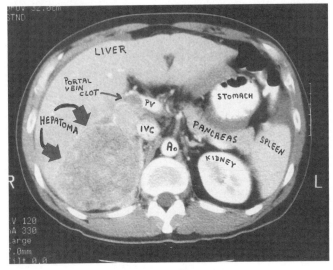

Figure 10.4. Hepatoma.

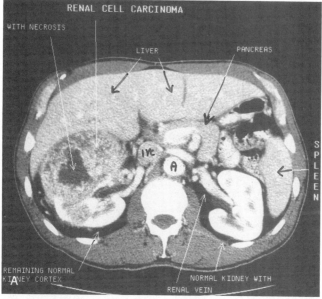

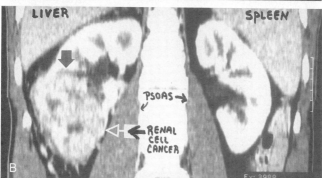

Figure 10.5.   A. Axial helical CT. B. Coronal CT reconstruction of renal cell carcinoma.

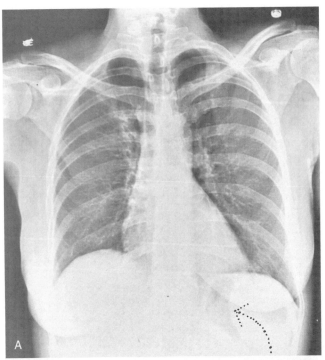

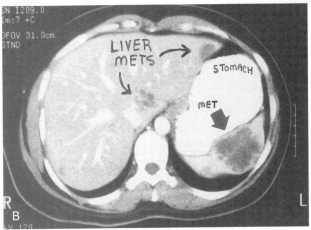

Figure 10.6. A. Medially-deviated stomach from splenomegaly. B. Same patient: splenic MET pushing on stomach.

Barrett's esophagus (metaplasia of lower esophagus—10% get adenocarcinoma), familial polyposis, undescended testicle, achalasia (lower esophageal sphincter fails to relax—squamous cell carcinoma precursor), adenomatous polyps in the colon, cirrhosis, chronic hepatitis, hemochromatosis, and multiple endocrine neoplasia (MEN) syndromes. Asbestosis, alcohol, tobacco, toxins, radiation, oncogenes, and certain viruses predispose to malignancy also.

CANCER CLINICAL PEARLS

- The most common intracranial neoplasms are metastases, gliomas, and meningiomas. (Metastases are more common than primary brain cancer in adults.)
- 95% of **brain metastases** are from lung, breast, GI, kidney, melanoma, or choriocarcinoma.
- **Multiple myeloma** is the most common primary bone malignancy in adults.
- There is relative hypertrophy of the left and caudate lobes in **cirrhosis,** which is at risk for hepatoma (Fig. 10.4).
- Hepatic adenoma may present with hemoperitoneum (classically, females on oral contraceptives).
- **Mediastinal nodal metastases** may come from the thyroid, head and neck, testicle, kidney, GI, lung, or breast.
- Lymph nodes smaller than 1 cm likely are benign, although different size criteria exist in different locations.
- The stomach is the most common extranodal site for Hodgkin lymphoma and the #1 site in the GI tract for non-Hodgkin lymphoma.
- Inflammatory or neoplastic invasion will usually blur fat planes or result in **"dirty fat"** (lighter fat than the usual near-black normal fat on computed tomography [CT]) (Fig. 10.3).
- Masses with fat in them include lipoma, liposarcoma, teratoma, dermoid, renal angiomyolipoma, lung hamartomas, and adrenal myelolipomas. This fat can be diagnosed with CT numbers.
- **Pancreatic carcinoma** has a dismal prognosis, regardless of resectability.
- Extramedullary hematopoiesis can cause a thoracic posterior mediastinal mass in anemics.
- **Hypervascular tumors** or metastases include renal cell carcinoma, thyroid carcinoma, choriocarcinoma, melanoma, hepatoma, islet cell tumors of the pancreas, and some breast cancers.

- Metastases in the liver are 20 times more common than hepatomas.
- Calcified liver metastases are typical for adenocarcinoma of the colon (and necrotic metastases).
- **Pseudomyxoma peritonei** is a belly full of jelly. Ovarian cancer or appendiceal mucinous cystadenocarcinoma can completely fill a bulging abdomen with loculated cystic collections of gelatinous ascites and masses.
- **Intraperitoneal tumor spread** occurs in ovary and GI tumors (colon, pancreas, stomach), and may involve the omentum (omental cake); or "drop metastases" may land in: 1) top of the sigmoid, 2) front of the rectosigmoid, 3) right paracolic gutter, and 4) ileocecal regions.
- **Carcinoid** is the most common primary neoplasm of the small bowel. **Rule of $\frac{1}{3}$s** says: $\frac{1}{3}$ metastasize, $\frac{1}{3}$ are multiple, and $\frac{1}{3}$ have another malignancy. Carcinoid syndrome (postmeal diarrhea, wheezing, sweating, and rarely flushing) requires liver metastases (bypasses the portal vein and, thus, serotonin is not immediately filtered or processed by the liver) or a pulmonary primary carcinoid.
- **Pheochromocytoma: rule of 10s** says: 10% are outside the adrenals, 10% are bilateral, and 10% have metastases.
- Double duct sign = dilated pancreatic duct and common bile duct indicating pancreatic carcinoma.
- Cushing's syndrome of elevated glucocorticoid production is usually (70%) due to adrenal hyperplasia from a pituitary adenoma or ectopic ACTH secretion.

STAGING

- **Colon cancer:** Dukes A = mucosal only, B = extends through wall, C = lymph node metastases, D = distant metastases.
- **Breast cancer:** I = less than 2 cm; II = 2 to 5 cm or positive axillary nodes (palpable but not matted or fixed); III = greater than 5 cm or fixed, matted nodes; IV = distant metastases or supraclavicular nodes.
- TNM **lung cancer** staging tricks: T1: smaller than 3 cm; T2: larger than 3 cm; T3: less than 2 cm from carina, with effusion or pleura; T4: invades organs; N1: same side hilar nodes; N2: same side mediastinal nodes; N3: opposite side nodes; M1: metastases.
- **Simplified lung cancer** stages: I = no nodes and not other

stages, II = NI, IIIA = T3 or N2 (difficult surgery), IIIB = T4 or N3 (unresectable, radiation threshold), IV = metastases (chemotherapy threshold).
- **Prostate cancer:** 1 = not palpable; 2 = palpable; 3 = extension through capsule; 4 = fixed to pelvic wall, invasive, or metastases.
- **Hodgkin lymphoma:** 1 = organ of origin, 2 = same side diaphragm, 3 = both sides diaphragm, 4 = metastases/extralymphatic.

Suggested Readings

Dahnert W. Radiology review manual, 2nd ed. Baltimore: Williams & Wilkins, 1993.

Ravin CE, Cooper C, Leder RA, eds. Review of radiology, 2nd ed. Philadelphia: WB Saunders, 1994.

Chapter 11

Horizons in Radiology

Helical computed tomography (CT) is a new type of CT scanner that can acquire data in a continuous fashion by moving the patient constantly through the doughnut while the thin x-ray beam spins around the patient. This creates a spiral of two-dimensional data that can be manipulated later to look in any plane (multi-planar reconstruction), with overlapping slices, or even in three dimensions (3-D) (Figs. 11.1–11.3). Conventional CTs take successive axial pictures of contiguous slices with no overlap.

Helical CT improves identification and characterization of small structures like pulmonary nodules which could have been missed using conventional axial slice CT (due to breathing differences between slices). The ability to reconstruct overlapping images may also help identify common bile duct, renal, or ureter stones. The **multiplanar and 3-D reconstructions** can help stage pancreatic carcinoma (to evaluate for vascular encasement that makes the patient inoperable); help visualize abdominal aortic aneurysms or renal artery stenosis (see Fig. 11.8); and can better depict facial, cervical, or orbital fractures. The 3-D imaging may also be helpful in CT angiography. Flight simulators hooked up to CT scanners allow for virtual reality diagnostic flights through the colon, sinuses, or tracheobronchial tree (**virtual endoscopy**) (Fig. 11.4).

Digital and computed radiography may make film hard copies obsolete in the near future. The imaging data is stored digitally rather than on conventional analog film. This image can then be postprocessed or rewindowed (similar to CT windows) to look for specific structures. For example, a chest x-ray can be evaluated with bone windows (to look for rib fracture), soft tissue windows (to look for foreign objects or gas), or tube windows (to look for the course of tubes in the ICU patient) (Figs. 11.5 and 11.7). Portable cross-table lateral C-spine films may be rewindowed to look at C7–T1 (Fig. 11.6). Digital information also lends itself to

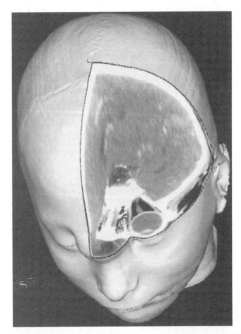

Figure 11.1. Three-dimensional CT (helical) of frontal encephalocele.

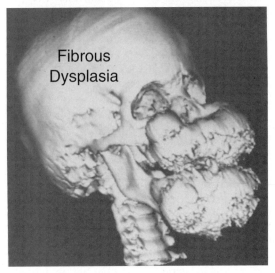

Figure 11.2. Three-dimensional CT.

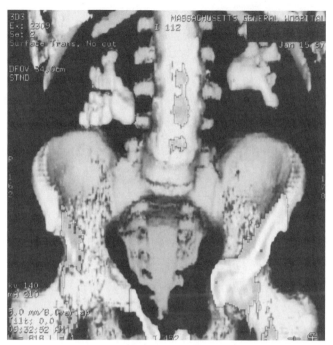

Figure 11.3.   Three dimensional helical CT of bones and bilateral staghorn calculi.

teleradiology, data transmission, and multiple viewing sites. One drawback of digital imaging is artificial contrast and edge enhancement. While these changes may make an underexposed portable film more readable, they also may make the interstitial markings falsely prominent, simulating mild heart failure or interstitial disease.

Advances in **magnetic resonance (MR)** have made it easier and quicker to obtain an MR of just about anything. Musculoskeletal MR is widely used in the knee, shoulder, and spine, as well as for soft tissue tumors. Newer MR applications in joints include the ankle, wrist, elbow, and even temporomandibular joint. **Dynamic renal or liver MR** looks at the effects of contrast over time to better characterize lesions, thus decreasing the necessity for biopsy in many cases. Organ-specific contrast agents in MR are

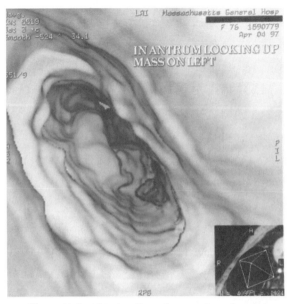

Figure 11.4. Virtual reality endoscopy stomach mass.

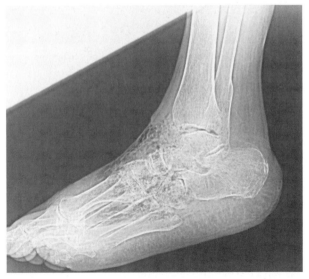

Figure 11.5. Gas gangrene digital film with soft-tissue windows.

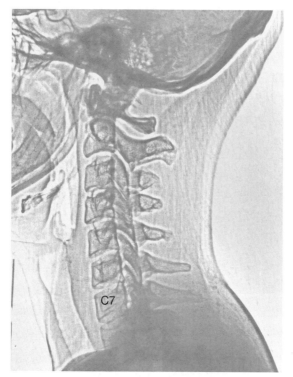

Figure 11.6. Inverted digital C-spine. C7–T1 are easier to see.

being developed. "Open MR" (shaped like an arc instead of a circle—for claustrophobia), near real-time CT, and portable CT are now available for interventional procedures like biopsies. **Functional MR** is a new window into brain function and exquisitely examines relative blood flow to various areas of the brain. Applications in neurology, psychiatry, and neurosurgery are being developed. Diffusion and perfusion MR may identify ischemic brain at risk for infarct. Thrombolytics may be given in the acute setting of some "brain attacks."

   Interventional radiology has an arsenal of new ways of treating and diagnosing diseases that are constantly unfolding, including peripheral angioplasty, carotid and cerebral angioplasty, gas-

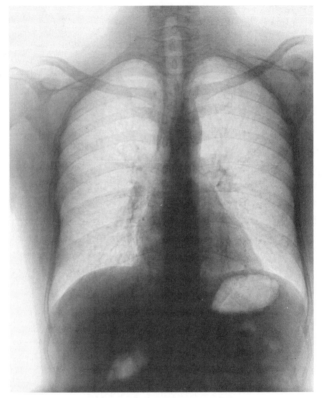

Figure 11.7. Inverted digital film.

tric ulcer, biliary and intravascular stents, selective embolizations, percutaneous repair of abdominal aortic aneurysms, and percutaneous biopsy or drainage of just about anything. Radiofrequency and alcohol ablation show promise for liver tumor therapy. Cooperative efforts with surgeons and endoscopists may foretell a multidisciplinary approach to minimally invasive therapies. Real-time imaging guidance for surgery is already happening.

Nuclear medicine offers disease and organ-specific studies to look for colon or ovary cancer recurrence, or to treat bone-pain from metastases. Monoclonal antibody-directed imaging and therapy may hold promise.

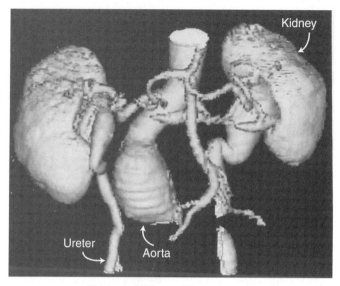

Figure 11.8. Three-dimensional CT.

Radiology is an exciting, dynamic and challenging field, but should not be foreign to today's technology-overwhelmed student. It can be very rewarding to make an important finding on your patient while on-call or in the emergency department. It may seem hard at first to grasp the subtleties of radiographic interpretation. Be persistent, boil it down to the basics, and take small bites. Remembering a few basic principles will make your nights run a little smoother. If you think you are overwhelmed or scared, remember how the patient must feel.

Index

Page numbers followed by t or f indicate tables or figures, respectively.